Praise for Debra Haffner's books

FROM DIAPERS TO DATING
A Parent's Guide to Raising Sexually Healthy Children

"A valuable book...to develop a comfortable, ongoing constructive rapport with your children about sex." *—Washington Post*

"Realistic, practical, and informative—the best kind of guide for being a better parent...Haffner [offers] a clear-eyed assist with deciding what sexual values to impart to children, and then advice on coupling those values with accurate, age-appropriate information." *—Kirkus Reviews*

"A terrific book." *—Time* magazine

"This book is an invaluable resource that you'll come back to time after time. Filled with sound, sensible advice and guidelines, it covers everything from what to do when you find your kids playing doctor to how to respond when your preteen wants to read Playboy. Not only does it teach you how to talk to your kids about sexuality, it also helps you decide what you want to say to them."
—Lynda Madaras, author, *The "What's Happening to My Body?" Books for Boys and Girls*

"I think every household with children under sixteen should have this book!...Debra Haffner is just the right mix of parent and friend; she defies labels of conservative or liberal. Rather, she is an advocate for children and for their right to have good rules and reliable knowledge so that their sexual and emotional growth is nurtured and safeguarded."
—Dr. Pepper Schwartz, Professor of Sociology, University of Washington

from DIAPERS *to* DATING

A Parent's Guide to Raising Sexually Healthy Children

REVISED SECOND EDITION

DEBRA W. HAFFNER

NEWMARKET PRESS

New York

This book is published in the United States of America and Canada.

Revised Second Edition

10 9 8 7 6 5 4 3 2 1

Library of Congress Cataloging-in-Publication Data
    Haffner, Debra,
        From diapers to dating : a parent's guide to raising sexually healthy children / Debra W. Haffner. —2nd ed.
            p. cm.
    Includes bibliographical references and index.
    ISBN 1-55704-623-9 (pbk. : alk. paper)
    1. Sex instruction for children. I. Title.
    HQ57.H343 2004
       649'.65—dc22

                                        2004004760

ISBN 978-1-55704-810-3

Quantity Purchases
Companies, professional groups, clubs, and other organizations may qualify for special terms when ordering quantities of this title. For information, write Special Sales Department, Newmarket Press, 18 East 48th Street, New York, NY 10017; call (212) 832-3575; fax (212) 832-3629; or email info@newmarketpress.com.

www.newmarketpress.com

Designed by Timothy Shaner

Manufactured in the United States of America

Table of Contents

Contents

With overwhelming love and appreciation to my two best teachers: Alyssa Haffner Tartaglione and Gregory Joseph Haffner Tartaglione. And to the man who makes everything possible, Ralph Tartaglione.

Acknowledgments

This book has been a joy to write! Although I did most of my writing before the sun even came up, I felt supported by a large number of people as I wrote.

I am very grateful to both the staff and the board of directors at SIECUS for their encouragement and support of the first edition. I have adapted the messages for children found in the boxes in each chapter with permission from two excellent SIECUS publications: "Guidelines for Comprehensive Sexuality Education, Kindergarten–Twelfth Grade" and "Right from the Start: Guidelines for Sexuality Issues, Birth to Five Years."

I am grateful to Amy Levine, who conducted much of the painstaking research for this book. She was always eager and willing to track down obscure references, arcane facts, and little details.

I am grateful to my friends and colleagues for sharing their stories of raising their children with me. I have changed the names of their children to protect their privacy. In particular, I would like to thank Ilene Wachs, Linda Bearinger, Dr. Elizabeth Feldman, Dr. Scott Spear, Barbara Levi-Berliner, Jodi Wallace, and Dr. Pepper Schwartz for sharing their stories with me. I am also grateful to the thousands of parents who shared their stories with me during the five years since this book first came out.

I am indebted to my colleagues and dear friends who read drafts of this manuscript: Shannon Colestock, Dr. Rosella Fanelli, Barbara Levi-Berliner, Dr. Doug Kirby, Dr. Bob Selverstone, Dr. Pepper Schwartz, and Monica Rodriguez. Their comments and suggestions made this a much better book.

On a personal note, I am so thankful to Dr. Pepper Schwartz for convincing and teaching me that I have books inside me to write.

This book was born at a Renaissance Weekend where I first met Esther Margolis, the publisher of Newmarket Press. Thank you to Phil and Linda Lader for including me in these weekends and the worlds they have opened up to me.

Thank you to Esther for believing in the importance of this project and for her unfailing support. And this is a much better book because of the loving care of my editor, Elissa Altman, at Newmarket Press.

I want to thank my parents, Harriet Haffner Hetherington and Saul Haffner, for the excellent sexuality education I received growing up in their home. Those early dinner table discussions are the foundations of this book. I am so lucky to have them as my parents and as my friends.

This book is dedicated to my husband, Ralph Tartaglione, and our two children, Alyssa and Gregory. All three of them are my daily teachers, my support, and the loves of my life. This book would never have happened without them.

Foreword to the Revised Second Edition

What a difference a decade makes. In 1998, when I wrote the first edition of *From Diapers to Dating*, we had not heard of MRSA, West Nile Virus, or al Qaeda, and few of us could have imagined September 11th. Our world seems both smaller and less safe than it did when I wrote the first edition of this book.

Influences on our sexuality have changed as well, necessitating an update of this book. Headlines about out of control teenagers to the contrary, the reality is that young people are actually more sexually responsible than they were fifteen years ago. Teenage pregnancies in the United States are at their lowest, dropping steadily between 1990 and 2005. Teen birth and abortion rates dropped during the same period. Teen alcohol and drug use rates also decreased.

In 1998, I wrote a small section on the Internet and children; as instant messaging, chat rooms, social networking sites, and increasingly explicit sexual materials have become more available through the Internet into our own homes, it was time to expand this section. It seems that every week we learn of adults who have been arrested in Internet stings because of their use of Internet child pornography and attempts to meet children offline. It is both more difficult and more critical to help keep our children safe from those who would hurt them. But, the growth of the Internet also means that there are ever more resources for parents to help them talk with their children about sexuality—in the first edition, I listed only a few Web sites; this edition includes many more.

Changing reproductive technologies makes the answer to "Where did I come from?" increasingly complex for some parents. Twenty-five years ago, the first test tube baby was born; today, more than 35,000 babies are born each year as a result of assisted reproductive technologies, and surrogate arrangements are more common. The section "When Your Child Isn't Conceived the 'When Mommy and Daddy Love Each Other' Way" has been expanded to include these new issues.

Our children are being exposed to issues about sexual orientation at a younger age. Today many prime-time TV shows now have a gay or lesbian character. The *New York Times* and nearly two hundred other newspapers now include announcements of same-sex unions. Same-sex marriages are now legal in Massachusetts, as are civil unions in several other states. Recent court decisions have upheld some of the rights of gays and lesbians. The need to educate our children about these issues and our own family values has become more pressing.

Our understanding of childhood gender-variant behavior has increased, and new organizations and approaches have emerged to help children who don't quite fit into a neat binary male/female gender system. And the challenges posed by the intersexual rights movement have also increased. Formerly known as hermaphrodites, adults who were born with ambiguous genitalia are now questioning gender assignment at birth and insisting on being called intersexuals. The transgender movement has become more vocal and visible.

On a more trivial level, popular culture has continued to change. The Spice Girls have been replaced by Paris Hilton and Lindsay Lohan. Britney Spears is all grown up now with children of her own, but she is still revered by preteen girls who want to emulate her scanty dress and strutting mannerisms. Baggy pants and jeans have been replaced by midriff tops and tight, low-slung jeans.

My own perspective has also grown and changed in the past

years, and these changes are also reflected in this new edition. I left my position as president and CEO of the Sexuality Information and Education Council of the United States to complete my preparation for the ministry, and in 2003, I was ordained as a Unitarian Universalist minister. My son, who was five when the first volume was published, is now in high school, and my daughter has finished college and is on her own. I've learned firsthand some of the challenges of raising a son instead of a daughter and helping a young woman and a young man navigate the changing world of adolescence.

I've also had the opportunity to speak to tens of thousands of parents since this book was first published. Many of them asked questions that were not included in the first edition. The question-and-answer sessions after my talks often challenged me, and I have tried to incorporate some of my learnings into this edition. I have also written a new book, *What Every 21st-Century Parent Needs to Know*, to talk about a wide range of parenting issues.

What hasn't changed, though, is the desire of parents to do a better job than their parents did in educating their children about sexuality. Indeed, if anything, I have noticed that parents feel more urgency in addressing these issues in a way consistent with their own values. We want to keep our children safe, but more, we want them to grow up into adults who can enjoy and appreciate their sexuality in caring intimate relationships. I hope that this book will help!

Foreword to the First Edition
by Alyssa Haffner Tartaglione, age 13

My mother wrote this book.

It hasn't always been easy having a mother who is a sexuality educator. Sometimes, it's hard to tell my new friends, their parents, and even my teachers what she does for a living.

But all in all, the positives outweigh the negatives. Ever since I can remember, we've talked about sexuality in our home—at the dinner table, watching TV, and now, in the car.

Here's some of what's been good:

I always can come to my mother with any question or problem.

I was *totally* prepared for puberty.

I've been a feminist since preschool. In fact, my birth announcements called me a "baby woman"!

She's great for science projects and homework. My first-grade science fair project was about kids and AIDS.

I'm more comfortable with my body than most of my friends.

No boy will ever be able to pressure me into something I don't want.

I know that as I get older, my mom will be there for me and my friends.

It's too bad that I don't know many kids who have this kind of relationship with their parents. Talking about sexuality with your kids is important. I know my mom has helped me and my brother grow up sexually healthy—let her help you!

Chapter 1
The Basics

I *thought it would be easy.*

After all, I am a sexuality educator with more than twenty years of experience. I knew what I wanted to teach my children about sexuality; I knew the messages I wanted to give them; and I thought I had the answers to any questions that might arise. After all, I had conducted workshops for parents on communicating with their children about sexuality for years.

I was wrong.

Teaching my own children about sexuality is often more difficult than I thought. But it has also proved to be one of the most rewarding parts of parenting.

To be honest, I wasn't prepared when my less-than-two-year-old daughter shouted, "Vulva!" as she pointed to a Georgia O'Keefe painting in a museum. I wasn't prepared to handle my son's newly circumcised penis several times a day. I wasn't sure what to say when my four-year-old daughter developed an intense fondness for Barbie dolls. I didn't know what to do when Gregory at age three stopped using the bathroom at school.

And, as my daughter Alyssa became an adolescent, I was humbled almost daily as I tried to prepare her for the challenges she faced. In the middle of an argument at age twelve, she summed up the difference between theory and practice, "But Mom, you're *supposed* to be an expert in my age group!"

Every sexuality educator will tell you that parents are the first—and the most important—sexuality educators of their children. They provide children with their first understandings of gender roles, relationships, and values, and their first sense of self-esteem and caring. Parents educate infants and toddlers about their sexuality when they talk to them, dress them, cuddle them, and play with them. Older children continue to learn about sexuality as they develop relationships within their families and observe the interactions around them. Most young people want their parents to be their most important source of information about sexuality.

And most of us let down our children in this important area. More than eight in ten parents believe it is their job to provide sexuality education to their children, yet few actually do it. Only one in four adults in the United States report that they learned about sexuality from *their* parents, but most say they want to do a better job. Baby boomers appear especially committed to doing a better job than their parents. But few seem to have the skills, comfort, or information they need.

In fact, many of today's parents put off talking about sexuality until their child approaches puberty. Every week, I get calls from parents of children who are twelve or thirteen and they ask me what they should include in the "Big Talk." The problem is that the Big Talk doesn't work. It probably never worked. If your parents gave *you* the Big Talk, what you may recall is that your mother and father were embarrassed and uncomfortable. And, you probably let them off the hook as quickly as possible with a quip like, "Aw, Dad, I already know all of that."

And bad advice abounds in this area. Articles are written about how to give your children the Big Talk. For example, one column in *Family Life* magazine advised the mother of an eleven-year-old daughter to "have some small occasional conversations about sex. Have them in the dark or while you are driving someplace so you won't have to make eye contact. Sex talks can be embarrassing."

The reality is that sharing your values and educating your children about sexuality can be one of the joys of parenting. Yes, it can be difficult, especially if your parents didn't talk with you. But it doesn't have to happen in the dark and it doesn't have to be embarrassing. It can even be fun!

Several years ago, I went to a baby shower for a good friend who lives in another city. I met one of her sophisticated, urbane friends, who asked me what I did for a living. When I told her, she grimaced and said, "I don't believe in talking to children before they are ready." I politely asked her the ages of her children; she had a baby and a six-year-old son. I asked if her son had ever asked her where babies come from—a question most parents will hear from their children somewhere between ages three and five. She said, "Yes, and I told him he was born in a cabbage patch."

I was stunned. Here was a woman, my age, who had not only not answered her son's question but had lied to him as well. And she felt good about it. Something was wrong.

What had she taught her son about sexuality? Inadvertently, she had told him she wouldn't talk to him about it. When he eventually learns the truth—maybe just a few weeks later from an older schoolmate—he will also realize that she lies to him about important questions. Unfortunately, she had also told him that she discounted his curiosity. And, most importantly, she had told him she was not a good source for reliable information about sexuality. And when he does reach puberty, he will probably not want to talk to her about these issues even though *she* may finally be ready to talk with him.

I wrote this book because I believe that it is parents' job to give their children information about sexuality and their own family's values. Sexuality education actually begins in infancy, and there are important messages that your children need to know at each stage of their lives.

I also wrote this book because I know that parents have questions. They want to get their child started right; they want their child to grow up feeling good about their body and with the ability to have a positive lifelong adult relationship. They want to do a better job than their parents did in handling sexuality issues, but they aren't sure how to begin.

Sexuality education actually begins in the delivery room. It may be difficult to think of your child as a sexual being, but most professionals believe that sexuality is part of human beings from birth to death. Now, before you get upset, let me explain that sexuality is different than "sex" or "sexual behaviors." Sexuality is about who we are as men and women, not about what we do with a part of our bodies.

You see, sexuality is about much more than sex. People often hear only the "sex" part of "sexuality"; as a parent once told one of my colleagues, "You say sexuality, but we don't hear the *-uality*." Teaching your children about sexuality is not just teaching them about anatomy and reproduction. It is teaching them who they are as boys and girls and laying the foundation for who they will grow up to be as men and women. It's about giving them the skills to develop good interpersonal relationships, now and in the future.

The organization I used to head, the Sexuality Information and Education Council of the United States (SIECUS) says sexuality encompasses an individual's sexual knowledge, beliefs, attitudes, values, and behaviors. In other words, your sexuality is not just shaped by your body and your feelings. It is also shaped by your cultural background, your family history, your education, your experiences, and your religion.

Sexually Healthy Families

I believe that sexually healthy families raise sexually healthy children who grow up to become sexually healthy adults. Sexually healthy children feel good about their bodies; are respectful of family members, other children, and other adults; understand the concept of privacy; make age-appropriate decisions; feel comfortable asking their parents questions about sexuality; and are prepared for the changes of puberty. In a sexually healthy family, parents consider educating their children about sexuality to be as important as teaching them about family responsibilities, religion, and self-esteem. They are "askable parents": Their children know they can approach them with their questions. They seek opportunities to talk with their children about sexuality rather than always waiting for questions. And they remember that what they *do* and how they treat each member of the family is more important than what they *say*.

Learning about sexuality is an ongoing process. Each year, sexually healthy families provide a foundation for what is to come. They do it little by little. Of course, parents should not give five-year-olds detailed information about contraceptives, but they can begin by teaching them that every baby needs love and caring and that people can plan the number of children in their families. This sets the stage for later, more in-depth discussions.

The Key: Finding Teachable Moments

Sexuality educators frequently talk about looking for "teachable moments" or the "golden opportunity." Rather than waiting to give your child that single Big Talk about sexuality, you look for those special times when you can easily bring up issues. If you see a pregnant woman in the park or store when you are with your four-year-old, you can say, "That woman is pregnant. A baby is developing inside a special place in her body called a uterus." When you are driving in your car with your nine-year-old and a story comes on the radio about new treatments for AIDS, you can talk a bit about your feelings about the AIDS epidemic. In each of the upcoming chapters, I'll help you consider how *you* personally can look for and use these teachable moments.

Today, children are continuously exposed to messages about sexuality. Television, movies, music, advertisements—and, most importantly, friends and acquaintances—all teach them from the very youngest ages about relationships, beauty, attractiveness, and yes, sex. And most of us do not like the messages they receive.

I believe that it is up to you and your partner to decide on the messages and values you want to give your child about sexuality. Throughout this book, you will find questions and exercises to help you decide what you want to teach your children. Every family has its own set of sexual values; it is your right and your responsibility to share them.

But do you know which values about sexuality you want to communicate? Many parents find it helpful to think about their sexual values in advance, talk to each other about them, and then to decide which messages they want to share. Although most people have a general feeling about their sexual values, they will often find them more complicated to discuss than they realize.

Here's a short quiz to help you start thinking about your own values about sexuality and your children. There are no right or wrong answers. The important thing is to think about your own beliefs and to determine the type of messages that you want to give to your children.

Do you agree (A) or disagree (D) with these statements?
_____ Boys and girls should have the same toys in their toy chests.
_____ I am comfortable having my child see me nude.
_____ Infants should be allowed to touch and enjoy their genitals.
_____ I wouldn't mind if my children were gay.
_____ It is the mom's job to teach about sexuality.
_____ Five-year-old twins of different genders can bathe together.
_____ Children need to know the correct names of their genitals.
_____ You can hurt children if you teach them about sex too early.
_____ Parents should never fight in front of their children.
_____ It is cute when seven-year-old girls have boyfriends.
_____ Parents can pierce their toddler girls' ears.
_____ Parents can pierce their toddler boys' ears.
_____ Cartoons contain too many sexist images.
_____ I want to be the one to teach my children about intercourse.
_____ You can spoil children by loving them too much.
_____ Parents should closely monitor children's television use.
_____ My eleven-year-old can go on group dates.
_____ Parents should set the standards for dress for children until high school.
_____ I want my child to wait until marriage to have sexual intercourse.

These are just a few of the issues you need to think—and talk—about with your family. They are difficult questions. You and your partner may not have the same answers to some of them. That's okay. Parents do sometimes have differing values about some of these issues. It is important to talk about these differences with each other, and then to discuss how to handle these disagreements with your children. For example, you may be pro-choice, and your partner may be pro-life on the controversial issue of abortion. Letting your child know that people can love each other even when they have differing points of view can be an important lesson. Throughout this book, both of you will have opportunities to think about your own values and what you both want to teach your children.

I will also try to provide you with the best information available on how to raise a child who will become a sexually healthy adult. I will differentiate between those issues that are based on an individual family's values, like your attitudes about premarital sex and contraception, and those that would interfere with your child's ability to grow up to be sexually healthy. For example, your family may want your child to know that your family believes that masturbation is wrong, but in order not to instill lifelong shame, it is critical that all children at puberty know that it does not cause physical or mental harm.

This book is about helping you give *your* children *your* messages. But I do have certain personal biases that I want you to know. I believe it is up to parents to set the foundation for their children's future sexual learning and, frankly, their future adult sexual lives. I also believe it is always wrong to lie to children. I feel strongly that teaching about sexuality is an ongoing process. You need to think through how you will educate your children about these important issues, and not just wait until your child asks you questions. I will also tell you my personal opinion about some of these issues and how we have handled them in our home—as a way to encourage you to think about how you want to handle them in yours.

I have been a sexuality educator for more than twenty-five years. For twelve years I was the president of SIECUS, the country's largest clearinghouse on sexuality education. I have been married for twenty-five years. My husband, Ralph Tartaglione, and I are the proud—very proud—parents of Alyssa, age twenty-two, and Gregory, age fourteen. Before I entered the ministry in 2003, I was an active member of my church, taught Sunday school, and was involved with parent activities at both of my children's schools. In other words, I am like most of you reading this book.

And, as you will read, my husband and I have struggled with many of these issues in our own home as we raise our children. I am fortunate and privileged, however, to have special training in sexuality education and also to have the opportunity to talk with hundreds of parents each year about these important issues.

Guidelines for Communication

Here are some basic guidelines for talking with children about sexuality that I have found helpful in my personal and professional life.

Remember that children want to talk with you about your values. Children want to talk with their parents about sexuality issues. They need to hear your point of view. They want you to help them with the facts about sexuality, but they also want to know what you *think* and *feel* about these important issues. For example, during the early elementary school years, you will want to tell them the facts of reproduction *and* you will want to tell them your values about sex and parenting outside of marriage. Children love to hear stories about what it was like when you were growing up and how you dealt with some of these situations. In survey after survey, teens tell researchers that they want to learn about sexuality from their parents. But they also tell them that their parents do not talk to them enough about sexuality issues.

Don't just wait for the questions. Some children are brimming with questions about sexuality; some never ask any. In our family, as a child Alyssa was always curious and forthright; her brother Gregory has been less likely to ask us questions. When children are young, parents don't wait until children ask to teach them to look both ways when they cross the street or to not to touch a hot stove. Nor do we wait until they ask us to teach them about God or our religious traditions. Some things are too important for them to know to wait until they ask. It is a parent's job to teach children about the world and about values. This includes teaching them about sexuality. It is up to you to decide what is important for your children to know about sexuality and then find ways to tell them.

Reward your children's questions. It is never a good idea to tell your children that they must wait until they are older before you will answer their questions. Let them know that you are an "askable parent." When they ask you a question, they are letting you know that they trust you to give them an honest answer. By saying, "I'm glad you asked me that," you are letting them know that you want to help them with difficult issues.

It's okay if you don't know the answer. Parents are often worried that they won't know how to answer their children's sexual questions. If you don't know an answer, say so. *Never make something up.* Your child may benefit by knowing that you are not invincible. Tell your children that you will look up the answer to their question. And be sure you do get back to them with the answer. If your child is school age, you might suggest that you can go to the library together and look up the answer or seek it on the Internet. The books and Internet sites listed in the appendix of this book can help you find answers.

In many ways, sexuality is an easier subject to discuss with children than other important life issues, such as religion or death. I was initially stumped about how to respond when my three-year-old asked, "Why do people have to die?" In many cases, children's sexuality questions (particularly about anatomy and reproduction) have factual answers that can be found in books.

In every chapter, I will give you age-appropriate messages about particular subjects. These messages have been adapted from guidelines that have been developed by leading educators, physicians, and youth workers who were members of the National Guidelines Task Force.

It's natural to feel uncomfortable. Parents often tell me that they are afraid that they will seem uncomfortable or show that they are embarrassed when they talk to their children about sexuality. That's okay. I am not always comfortable with all of the questions and situations that I face with my children about sexuality, and I do this for a living! Tell your child that this is hard for you, and if it is true, that your parents never talked to you about these issues. And then tell them that you want to talk to them about sexuality because you love them and want to help.

Parents of older children are sometimes faced with personal questions like "Mom, when did you first have sex?" "Dad, how often do you and Mom have sex?" "Mom, did you ever have sex with a woman?" and so on. It is perfectly acceptable not to give your own sexual history to your children; in fact, it is probably not a good idea. You can respond to these questions by saying, "I'm not comfortable sharing my own personal behavior with you now. But it sounds to me like you are wondering how people decide if they are ready for sex. Let's talk about that."

Find teachable moments. Sexuality educators define "teachable moments" as the times that arise naturally to easily provide sexuality information: You are watching a television program or movie and a sexuality issue comes up; you are reading a book together; you are giving your preschooler a bath or changing your toddler's diaper; you hear a story about sexual harassment on the radio when you and your fifth grader are together in the car. Each chapter in this book will help you identify and take advantage of these teachable moments.

Listen to your children. When your children do ask you a question, start by asking them what they already know or what prompted them to ask this question now. It is sometimes simpler as parents to talk *at* our children; it's often harder to really *listen* to them. Try hard to listen to the concerns of your growing children. Although your third grader's crush on a boy in her class may seem silly to you, it is very important to her. And your willingness to listen now, sets the stage for your talks when she is an adolescent and has to make decisions about such issues as dating and sexual behavior.

Facts are not enough. It is important to share information with your children, but it is not enough. You also need to share your feelings, attitudes, values, and beliefs with your children. Be sure to tell them *why* you feel the way you do. Telling them the "why" behind your values will help them think and will also teach them more about their own family, culture, and religion. Be assured that a vast amount of research tells us that children look to parents for help in developing their own values. Many parents worry that their children will develop values that are radically different than their own. You can relax about this exploration. Most American children grow up to adopt a value system fairly similar to their parents'.

It is also crucial that children from an early age understand that there is a difference between thoughts, feelings, and behaviors. Little children can be taught, for example, that everyone feels angry sometimes, but that it is never all right to hit another child or adult. Parents can help their older children understand that while it is natural to have all types of sexual thoughts and feelings, people are responsible for their own behaviors and should not act on such feelings if they are not consistent with their values.

Educate both your sons and daughters. Both boys and girls need information about sexuality, and they need similar education. Many families seem to be more comfortable educating their daughters than they are educating their sons. Some parents wonder if their daughters really need to know about masturbation or if their sons really need to know about menstruation. Throughout this book, I will stress giving similar information to children of both genders, except for a very few instances, where information is truly gender based. For example, both your daughter and son need to know about menstruation, but only your daughter needs to know the mechanics of sanitary protection.

Take the time to think about what you want to teach about sexuality. Trust me, you will face many of what SIECUS first labeled "Oh no, what do I do now?" situations. Almost every parent does: You find your child playing doctor; your child walks in on you and your partner making love; in the supermarket, your child asks you in a loud voice, "How come my penis gets hard?" It is helpful to think through your values *before* these situations arise and decide which messages you want to give. The exercises at the beginning of each chapter of this book will help.

You and your children's other parent need to talk about the messages and values you want to give your children about sexuality. (This is true whether you are coupled, married, remarried with children from both families, gay, divorced from your child's mother or father, or single but in touch with your child's other parent.) You may find out that you don't always agree. You grew up in a home where everyone walked around nude; he grew up in a home where the adults dressed in the bathroom and wore robes over their nightclothes. In your home, sex was talked about at the dinner table; in her home, sex wasn't talked about at all. Which messages do both of you want to give in your own home today about sexuality issues? And how will you handle disagreements about these messages? Do you believe it is important to present a "united front" on controversial issues? Or can your children learn that people who love each other don't always agree?

It is both parents' job to teach children about sexuality. Children need to learn about sexuality from both their mother and their father. In many homes, it seems that it is often the mother's job to talk about sexual issues. In other homes, it is gender segregated: Fathers talk to the boys and mothers talk to the girls. It is not unusual when I give a talk for parents that moms outnumber dads five to one. If you're a mom reading this book, suggest that your husband or partner read it as well.

It is helpful for children to learn about sexuality issues from both parents. In this way, they can learn that sexuality is a topic that is discussed openly in your home, and that men and women can both talk about these issues. And in single-parent homes or homes with gay parents, it can be helpful to ask for help from grandparents, friends, and other close relatives so that your child learns from both men and women.

Practice. As you read this book, think about how you will handle similar situations. If you are uncomfortable talking about sexuality or using certain words, practice with your partner or a good friend. I once trained a wonderful educator who was in her seventies when I hired her, who couldn't say the word "penis" without blushing. She went home one afternoon and spent an hour in front of a mirror saying "penis, penis, penis" until she could do so comfortably. If you think you'll die when it's time to talk to your child about how babies are made, practice the answer now. It will help when the teachable moment arises. Throughout this book, I give you sample answers to the really tough questions, like "How did the seed get in there?" or "What is gay?"

Use words and ideas that are appropriate for your child's level of development. Children are concrete thinkers. They don't develop the ability to think abstractly until sometime in adolescence. When your five-year-old asks you "Where did I come from?" he may be talking about geography, not sex. You may have heard the joke about the little boy who asks his father, "Daddy, where did I come from?" His father, seizing the teachable moment, launches into a long, detailed description of reproduction. The little boy interrupts, "No, Dad. Danny says he's from Cincinnati. Where do I come from?" That's why it's a good idea to always first find out what your child already knows: "Tell me what you know about where babies come from."

Finding out what your child is really asking is important at all of the ages discussed in this book. It is important to clarify your child's question. These problems in communication do not end in early childhood. One of my friends reports that one day when she and her husband were in the car, her eight-year-old daughter asked, "Mom, how do lesbians make love?" Flustered, she neglected to ask her daughter what she already knew and launched into a somewhat detailed, uncomfortable discussion about how two women might physically show each other affection. She finally paused for a breath and her daughter interrupted, "No, no mom. I meant, how can two lesbians have a baby?" My friend wistfully thought to herself, "Now, that's a question I know how to answer!"

Many parents are afraid that they will give their children too much information. For example, many parents have told me that they are worried that if they introduce the idea of intercourse, their child will want to try it. Relax—research shows that learning about sexuality does *not* cause young people to experiment. In fact, you will probably find out that if you do give them too much information, they will stop listening or start fidgeting. If you pay attention to your children's cues, you will know when you've said enough.

My colleague and friend Pamela Wilson, who is a sexuality educator, advises parents and teachers that "less is better than more." She tells parents to give children the simplest explanation and to move to more complicated information only if the child seems interested or asks more questions. If your four-year-old child sees a condom next to your bed and asks, "What's this, Dad?" try the simplest answer, "It's a condom." If, and only if, they ask, "What's it for?" do you say, "Your mom and I love you and your sister very much, but we've decided to not have any more children. The condom helps us do that." Most four-year-olds will be satisfied with that. Follow your child's lead.

It's okay to make a mistake. There have been several times when I have not handled a situation or an opportunity as well as I would have liked to—and I do this for a living! Don't worry if you make a mistake; you have lots of time to make up for them.

Let me give you an example: Three-year-old Alyssa and I were driving home from her day-care center, when she asked me, "How come I don't have a penis?" I answered, "Boys have penises and girls have vulvas." (So far, so good.) She went on, "Jason likes to touch his penis all day long." I was, to say the least, struck speechless, and yes, I hesitate to admit to you, I quickly changed the subject. By the time I got home, I was feeling pretty amused at my reaction but knew this wasn't the time to bring it up again. I decided to wait until the next time I bathed her and then I introduced the concept of privacy. (See "Genital Touching" in chapter 3 for more information on handling this.)

The point here is that I had time to clear things up later. So, if you say to yourself, "Well, I wish I had..." or "Boy, did I mess that up," just look for a time to reintroduce the topic later. And remember, it's okay to tell your child "I'm sorry" or "I was wrong." In fact, it teaches them that no one, not even their mom or dad, is perfect.

Remember that actions speak louder than words. When it comes to sexuality education, what we *do* is often more important than what we *say*. For example, we can tell our children that men and women should be equal, but if they see one partner constantly trying to dominate the other, those words are pretty meaningless. Or, we can tell them that their body is wonderful, but if we swat their hands away when they touch their genitals during a diaper change, we are teaching them that part of their body is bad. Most importantly, our relationship with our own partners sets the stage for our own children's intimacy in their adult years. (If you are a single parent, you will need to consider special issues that I will discuss in upcoming chapters.)

There is no such thing as too late. As you will read, I believe that sexuality education begins in infancy. But, if your child is eight or twelve, and you have not had a conversation yet, there is still plenty of time. It is almost never too late to begin. Remember, you have probably been teaching your child indirectly about sexuality issues for a long time. Try to remember how you handled some of these early experiences. Think about the messages you want to give now. And start looking for teachable moments. You might even want to tell your child, "I know I've never been comfortable talking about sex with you before, but I'd like us to be able to talk about these important issues now." Telling your child you are reading this book can be a teachable moment.

Understand that there is a difference between childhood sexuality and adult sexuality. Parents are often understandably upset when they see their children rubbing their genitals in public or find them playing doctor with the child next door. This is partially because parents often assign adult meanings to these behaviors. Is my child masturbating? Is my child normal? Is my child obsessed with sex? Most childhood sexual behavior is naive and curious. It is not usually directed to orgasm or erotic feelings as adults understand them. In each chapter in this book, I will give you information about what is expected sexual development at each stage of childhood.

Sexuality education is an ongoing process. Sexuality education cannot be reduced to the Big Talk. It is not an immunization that you give once or a few times in childhood. To raise sexually healthy children, parents must realize that, like other important value-related subjects, sexuality education is ongoing. We don't take our children to Sunday school once every few years; we know that teaching them about our religious traditions takes time and reinforcement. The same thing is true for sexuality issues: Giving them information throughout childhood reinforces that you want to talk to them about these important issues. It also allows you to tailor the information and values to their age and stage of development.

Don't forget to talk about the joys of sexuality. It is easy in today's world to focus on what is unhealthy about sexuality. We certainly don't want our children to face sexual abuse, pregnancy, sexually transmitted diseases (especially AIDS), or emotional trauma. Most parents don't want their children to grow up feeling afraid or guilty about sexuality. But when we start our conversations about sexuality with all of the don'ts, we may be teaching that all sexual feelings are negative or that all sexual behaviors have negative consequences.

Most parents want their children to grow up to become adults who understand that sexuality can be a wonderful part of life. We need to tell our children that loving relationships are often the best part of life and that intimacy is wonderful for adults. When you talk with your children about sexuality, you are telling them that you care about their happiness and well-being. You are also sharing your values. You are doing your job as a parent, and you are strengthening your relationship with your children: They are learning that they can trust you with this very important part of their lives.

Let's get started.

Communication Tips

- Remember that children want to talk with you about your values.

- Reward your children's questions.

- Don't just wait for the questions.

- It's okay if you don't know the answer.

- It's okay if you feel uncomfortable.

- Find teachable moments.

- Listen to your children.

- Facts are not enough.

- Educate both your sons and daughters.

- Take the time to think about what you want to teach about sexuality.

- It is both parents' job to teach children about sexuality.

- Practice.

- Use words and ideas that are appropriate for your child's level of development.

- It's okay to make a mistake.

- There is no such thing as too late.

- Remember that actions speak louder than words.

- Understand that there is a difference between childhood sexuality and adult sexuality.

- Sexuality education is an ongoing process.

- Don't forget to talk about the joys of sexuality.

Chapter 2
Infants and Toddlers
Birth to Age 2

Sexuality education for infants? Of course, infants do not need to know about reproduction, anatomy, and contraception. But they are beginning to learn about and discover their own sexuality, and parents are their most important teachers. As you talk to your children, as you hug them and kiss them, as you dress them, and as you play games with them, you are setting the stage for future sexual learning.

People sometimes laugh at me when I tell them that sexuality education begins in the delivery room. Of course, I am not talking about giving newborn babies facts about reproduction or sexual behaviors. But think for a minute: What was one of the first things you wanted to know about your baby: "Is it a boy or a girl?" And sexuality education about gender roles begins.

There is research that indicates that there are some innate differences between boy and girl babies. Studies indicate that boy babies may be more active and irritable than girl babies. On average, girl babies seem to reach such developmental

Values Exercise for Birth to Age 2

When my child finds his penis during a diaper change,
I will probably...

- ❏ a) move his hand and finish diapering him quickly.
- ❏ b) slap his hand and tell him, "That's nasty."
- ❏ c) smile and say, "What a wonderful body you have!" and continue diapering.
- ❏ d) let him play with his genitals until he stops and only then put the diaper back on.

When I buy clothes for my infant daughter, I will probably...

- ❏ a) dress her only in pinks and pastels.
- ❏ b) dress her only in primary colors.
- ❏ c) dress her in all colors, including her big brother's hand-me-downs.
- ❏ d) keep her in T-shirts and diapers and avoid the issue.

It is important for...

- ❏ a) the mother to provide the primary care to a new infant.
- ❏ b) the father to provide the primary care to a new infant.
- ❏ c) the father to help the mother provide care for the infant.
- ❏ d) both parents to share infant care equally except for breast-feeding.

A crying baby who has been fed and diapered should be...

- ❏ a) left to cry and fall asleep on his own.
- ❏ b) picked up immediately and rocked until the crying stops.
- ❏ c) briefly cuddled and put down.
- ❏ d) carried in an infant carrier all day.

milestones as sitting up, reaching for objects, and talking earlier than boy babies.

But boys and girls are also socialized differently from their earliest moments of life. In fact, our expectations for them as males and females may start before birth.

Mothers who know the sex of their fetuses in utero even describe fetal movement differently. Pregnant mothers of boys say that the fetus's movements are strong and vigorous while pregnant mothers of girls describe them as lively or moderate. When pregnant women do not know the sex of the fetus, they use words that are not gender linked. But studies show that male fetuses are not really more active than female fetuses. Gender stereotypes affect perception even before a baby is born!

Many people prepare differently for their new baby based on whether they know if the child is a boy or a girl. Parents-to-be select different wallpaper and accessories for the nursery, different colored infant clothes, and different announcement cards. Even new-baby congratulations cards are rigidly divided into pink or blue. Pastel yellow and green are available for those who want to buy clothes or paper the nursery before for the child's birth and don't know the sex; in fact, that's all that's often available if they don't know. Today, even newborn disposable diapers are coded blue and pink and feature different characters, although throughout history, genderless diapers did just fine.

However, with today's frequent use of amniocentesis and ultrasound, many of us *do* know the biological sex of our children months before their birth. And for some of us, it is during the middle of our pregnancies when we begin to think about how we want to raise that boy or girl child. Although it may be politically incorrect to say so, most of us secretly hope for a child of one sex or the other. In fact, surveys of people about to become parents in the United States have found that, for many of us, the ideal birth order is first a son, then a daughter.

We all know families who keep "trying for a boy" after their first three babies are girls. And I know women who have only wanted daughters. In fact, a friend of mine once spent three months in counseling prior to the birth of her child because she had been told she was having a son, and she couldn't imagine she would raise a boy well. To her surprise and chagrin, the doctor sheepishly told her immediately after the delivery that the technician must have misread the sonogram. She had had a girl!

I can relate to this story. At thirty-nine, I was pregnant again. As we were heading to the hospital for my amniocentesis, my husband and I debated whether we wanted to know the sex of the fetus. I did; he wasn't sure.

As I lay on the table, the technician moved the sound instrument around my expanding stomach. She asked us, "Do you want to know the sex?" Just then she moved the doppler so that it showed the place between the fetus's legs. My husband laughed, "I think we know what the sex is." Sure enough, our sixteen-week-old fetus was displaying a fully erect penis. We were having a boy! We were given several pictures of the fetus to take home, including one that quite explicitly showed our developing male child.

And then we began to worry. We were successfully raising our daughter, Alyssa. She was bright, funny, and definitely a developing feminist. We had bought her dolls and trucks, enrolled her for both dance and karate classes. We were raising her to believe that girls can do—and be—anything. We had read "Free to Be You and Me" to her and made sure she had nonsexist playthings and books. We had even sent birth announcements proclaiming, "It's a baby woman." But it wasn't going to be so easy with a boy. What if he wanted ballet lessons? (He didn't.) What were the masculine images we wanted to communicate? Should we pass down his sister's Barbie dolls? (We did.) Would we be good parents to a boy? (I think we've done fine.) And could we really send birth announcements announcing "It's a Sensitive New Age Guy?" (In case you are wondering, we did not.)

Studies show that parents often talk differently to boy infants and girl infants. In fact, many parents hold, play with, and even touch their girl babies differently than the way they do their baby boys. Some studies have even found that parents' views of their new babies differ by sex during the first 24 hours of life. Parents rate their new girl babies as more delicate, finer, and softer than they do their boy babies; newborn sons are rated as stronger, firmer, and hardier than newborn daughters.

During the first year of life, parents continue to treat their sons and daughters differently. Studies show that mothers look at their daughters more than they do their sons, hold them more, touch them more, and cuddle them more. Mothers are more emotionally expressive to daughters: They smile at them more, talk to them more, and respond to their needs more quickly. Even in the first few months of life, these behaviors encourage girls to be more social and more emotional than boys. Mothers wait longer to tend to their sons' needs, which perhaps gives them a greater beginning sense of autonomy and independence. Fathers play with and talk to their infant sons more than they do their infant daughters, and many fathers adopt a rougher kind of horseplay with their sons than they do their daughters.

During the first eighteen months of life, children themselves are also beginning to learn the differences between males and females. Babies pick out male and female voices as early as six months of age. Between a year and eighteen months, babies look longer at photographs of people of their own sex than they do of people of the other sex. Studies show that most children by the age of three can identify both photographs and dolls as male or female and can tell an adult whether they are a boy or a girl.

Sexual Development of Infants

It may be difficult to believe that newborn babies develop sexually. But in some ways, the first eighteen months of life are one of the most important times for learning about love and touch, and developing a sense of trust in the world. It is during their infancy that babies learn they are loved and how to love when we kiss them, hug them, and talk to them. Babies also learn if care providers will meet their needs or if they will have to tough it out. They learn quickly if they can count on their care providers to respond to their basic needs.

Actually, biological sexual development begins during pregnancy. Sexologists used to say that human beings were sexual from birth to death; new ultrasound technology has now verified that the sexual response system begins to develop in males during the middle stages of gestation. Erectile response begins to appear at around sixteen weeks; respiratory function does not begin until almost twelve weeks later. And although sexual response in female fetuses is not readily observable, it is assumed that the capacity for lubrication begins at this time as well.

During the first few months of life, infants begin to discover their bodies. By seven or eight months, they discover their hands and toes. About the same time, boys discover their penises. Girls, on average, seem to discover their vulvas about two months later. Infants love to put their fingers and toes in their mouth, and they love to touch their genitals. It may surprise you to know that baby boys have erections regularly throughout the day and during sleep, as many as three or more times a night. In fact, infant boys may become erect simply by crying, coughing, stretching, or urinating. Scientists believe that baby girls' vaginas lubricate about just as often, although for obvious reasons, this isn't easily observable and these studies have not been done.

Parents' Feelings

Although many people don't talk about this, taking care of an infant is often very sensuous for parents. There is nothing quite like the smell of a new, freshly diapered baby. The mixture of their own sweet smell coupled with the Ivory Snow–like fragrance of their clothes, the Johnson's Baby Shampoo, and the diaper lotion is unmatchable. I never understood until I had a baby why people wanted to hold other people's infants; now, as my children have grown up, I love holding other people's babies. And one of my favorite parts of ministry is leading child dedication ceremonies.

Breast-feeding can be a very sensuous and pleasurable experience. Some breast-feeding mothers are anxious that they have sexual feelings when they breast-feed their babies. Some even reach orgasm during breast-feeding. This is a perfectly expected physiological response: Oxytocin, the hormone that triggers the letdown of milk when a baby suckles, is the same hormone that triggers orgasm. It does not mean that you are having sexual feelings towards your baby or having sex with the baby. Your body is simply responding to your breasts' being stimulated in this way.

Actually, breast-feeding may not be the only time that you may have some type of sensual feelings aroused by your child. Taking care of an infant is a very intimate experience: You stroke each other, cuddle them, kiss their neck, and stroke their back. Some professionals believe that these types of physical interaction even set the foundation for what may be erotically appealing when the child grows up. A fleeting sexual thought during these moments is nothing to worry about; you can acknowledge it to yourself and then reassure yourself that you would never act on this behavior. One of the hallmarks of a sexually healthy adult is being able to discriminate between sexual behaviors that are life enhancing and those that would be harmful to oneself and others. *Any kind of erotic touching between you and your baby is absolutely off limits*. If you are finding it difficult to control or handle these type of feelings, you need to seek out and talk with a mental-health professional.

I don't want this to discourage you from cuddling with your baby. Just recognize that there is a difference between these intimate feelings and erotic ones that you could act on. Cuddling and stroking babies helps them learn that their bodies feel good when they are touched and also helps cement the parent-child bond. In fact, research shows that parents should hold their babies as much as possible. A loving touch helps babies grow and develop. And when touch is violated, as in the horrifying cases of child abuse or parental neglect, the child may grow into an adult who may be incapable of having healthy adult sexual and intimate relationships.

The Importance of Touch

Some psychologists theorize that loving touch helps set the stage for adult intimacy.

Touching and holding children teaches them how we feel

about them. When parents and caregivers convey love and delight when they hold their infant, the infant learns they are loved. Conversely, if the adults are scared or tentative or uncomfortable, the infant seems to learn that they are not quite all right. These early messages about feelings and our bodies may persist throughout our life.

Developing a sense of trust in the world is an important foundation for adult mental health, including sexual health. Developmental psychologist Erik H. Erikson wrote that the first psychosocial crisis in life is to resolve "basic trust versus basic mistrust." According to Erikson, between birth and eighteen months of age, children learn whether their needs will be met and if they can trust the people and world around them. If their needs are met, they develop the ability for intimacy and a sense of hope. Erikson wrote that feeding is the primary way that an infant resolves these issues: Can I trust that they will feed me when I am hungry?

But many other psychologists believe that touching and holding are just as important: When I'm crying, will I be picked up? Can I count on you to show me that you love me? Responding to your infants' needs teaches them that they are loved and that the world is a predictable, safe place. It boosts their self-confidence and their ability to believe that when they need help and support, they will get it.

Infants learn about their world and their bodies through touch. In fact, in extreme cases, babies who are not touched may develop a disease called "failure to thrive" and may die. They actually wither away: They don't eat, they don't absorb nourishment, and they withdraw from the world. (Not all cases of failure to thrive are related to an absence of touching; some babies do have an inability to take nourishment. If you have any concerns, see your pediatrician.)

In the early 1960s, two researchers did a famous experiment with rhesus monkeys. They separated some monkeys from their mothers at birth and offered them "surrogate" wire mothers to hold. Some of the monkeys were only wire frames; others were covered with soft fabric. The monkeys would cling to a dummy covered with soft fabric, but rejected one made up just of wire, even when it had a bottle of milk attached. In other words, the baby monkeys preferred a soft touch to being fed when that comfort wasn't there! These baby monkeys denied touch also grew up to be troubled adult monkeys. They were more likely to bite and scratch; the males didn't approach females sexually; and the females were mostly infertile. Some psychologists believe that these studies show that a lack of touch in infancy can affect future adult sexual and intimate relationships!

The Circumcision Decision

Parents may be surprised how early some sexuality-related issues surface. It may seem as if the challenges of raising an infant rarely have anything to do with sexual issues—unless it is whether you, as the new parent of a screaming, nonsleeping baby, will ever have sexual relations with your partner again. (And that's a topic for another book!) But actually, many parents will face a sexuality-related decision about their infant son even before they leave the hospital: whether or not to circumcise him.

Circumcision is a tough decision today. Circumcision is removing the foreskin of the penis. At the turn of the century in the United States, only Jewish and Muslim parents circumcised their male infants. By the middle of the century, almost everyone did. But at the beginning of the twenty-first century, it is increasingly a matter of choice again. Today, approximately two-thirds of all male infants in the United States are circumcised. (Circumcision rates vary by region of the country: Only 40 percent of male babies in the West are circumcised compared to nearly 80 percent in the Midwest.)

There is a great deal of debate about circumcision today. At one time, medical professionals thought that circumcision was important for health or sanitary reasons. In 1999, the American Academy of Pediatrics (AAP) revised its recommendation on infant circumcision. As early as 1971, the Academy said that there was no absolute medical reason for every male infant to be circumcised. In 1999, after reviewing nearly 40 years of medical research, they said, "Circumcision is not essential to a child's well-being at birth, even though it does have some potential medical benefits. These benefits are not compelling enough to warrant the AAP to recommend routine newborn circumcision. Instead, we encourage parents to discuss the benefits and risks of circumcision with their pediatrician, and then make an informed decision about what is in the best interest of their child." The AAP recommends that if parents decide to circumcise, it is essential that pain relief be provided to the infant. Ask the doctor to use EMLA cream (a topical anesthetic), a dorsal-penile nerve block, or subcutaneous ring block; ice is not enough to relieve your baby's pain and stress. In the words of *Medem*'s editor-in-chief, Dr. Nancy W. Dickey, "The bottom line: Circumcision is an elective procedure."

Here are some reasons you might consider circumcision:

- You are Jewish or Muslim. Infant circumcision is part of these religious traditions.
- You don't want to have to teach your son to clean his foreskin.

- Circumcised babies have a lower incidence of urinary tract infections and no risk of foreskin infections.
- Some uncircumcised men have medical problems, such as painful intercourse or infections, requiring them to be circumcised during adulthood. This is much more painful and dangerous than newborn circumcision.
- Cancer of the penis, though extremely rare, is higher in uncircumcised men.
- Uncircumcised men may be more susceptible to sexually transmitted diseases, although a person's behavior is "far more important" according to the AAP.
- If your child's father is circumcised, you want your son's penis to look like his father's penis.

Here are some reasons you might not consider circumcision:

- Circumcision is not part of your cultural tradition.
- Circumcision is painful for infants.
- If your child's father isn't circumcised, you want your son's penis to look like his father's penis.
- Like any surgery, circumcision poses some risks. In very rare cases (less than two times in one thousand) there is a chance of infection or even damage to the penis.
- You think circumcision is not natural. Boys are born with penises with foreskins; their intact penises should be left alone. Some people believe circumcision is a form of genital mutilation.
- Circumcision is done without the infant's permission. This should be an adult choice.
- Your child is ill at birth. Circumcision should never be done on a sick or medically unstable infant.

If you decide not to circumcise your son, be sure you receive clear instructions on caring for the foreskin as part of daily hygiene. As your son gets older, he will need to learn these techniques as well.

First Days at Home

Circumcised or not, during your first days at home, you may be faced with a new sexuality-related challenge almost immediately. This is a little embarrassing to admit, but I was completely unprepared to handle newborn Gregory's newly circumcised penis. (Yes, we did circumcise him; as a woman from a Jewish background, I felt it was important to continue this family tradition.) And many parents, particularly dads, report that they are uncomfortable cleaning their daughter's vulva and labia during diaper changes.

Why was I uncomfortable? Some parents are surprised that their boy babies have erections during diaper changes and worry that they are stimulating them too much. I knew that boy babies have erections on a regular basis, so I wasn't concerned about that. (In fact, obstetricians and neonatologists report that many newborn boys have erections in the first minutes after birth.)

But to be honest, I had just never handled a penis quite so often. Here I was, wiping, applying ointment, and moving his penis to clean it eight to twelve times a day. And, frankly, I was uncomfortable. I turned to some of my friends who already had sons. Some of them admitted that they had been so uncomfortable that they had avoided thoroughly cleaning their son's genitals during the first month. I soon found out that fathers of infant girls often felt the same way about cleaning their daughters' labia during diaper changes.

I called our pediatrician for advice. She told me to relax, that it was important to clean Gregory's penis thoroughly before each new diaper, and most importantly, she reassured me that my feelings were normal. She reminded us that there was nothing erotic about handling his penis this way. She told me that in a few days, it probably wouldn't bother me as much. I did what she suggested: I cleaned him, put lotion on him, and re-diapered him, what seemed like dozens of times a day. Practice makes perfect: She was right. In a few weeks, I was a pro!

Teaching the Parts of the Body

Bath time and diaper changes are wonderful times to start teaching your child the parts of their body—*all* the parts of their body. Many parents play this game to teach body parts with their five- or six-month-old children, "Here is your nose, here's your tummy, here are your knees, here are your toes." In addition to teaching the names of these body parts, these parents may be inadvertently giving a beginning message about their willingness to address sexual issues. These parents may be communicating to their child that a third of the body has no name, and that this third is different from all of the other parts of the body. How much more sex positive it would be if parents could learn to say, calmly and without flinching, "Here's your nose, here's your tummy, here's your penis or vulva, here are your knees, here are your toes." The difference: All the parts of the body that the baby is exploring have a name, and Mom and Dad can speak about all of them.

Parents often ask me why I feel that it is so important to teach infants and toddlers the correct names of all of the important parts of the body. After all, they tell me, they learned "private parts," "down there," "charlie," and "ding dong" from their parents. In fact, one parent in a class for preschool parents once told me that the correct names actually sound "dirtier" than the euphemisms.

I sometimes ask parents, "What do you call the elbow in your house?" and then, "What do they call the elbow in the house next door?" I follow that by asking, "And if your child got hurt outside, would someone outside your family understand when she said, 'ouch, I hurt my elbow really badly'?" The point is that the genitals are the *only* part of the body we teach euphemisms for, and it may mean that if something happens to your child, they will not be able to communicate with anyone outside of your home.

I believe parents should treat all body parts equally. When you use euphemisms only for the genitals, you are giving your child a message that these parts of the body are uncomfortable or different. You may, without meaning to or realizing it, even introduce a sense of shame or guilt about this part of the body. These feelings sometimes persist into adulthood, making it difficult for grown men and women to be comfortable with their bodies and sexual feelings. You may also be affecting your child's ability to tell you about sexual abuse incidents accurately. (See page 82 for more on helping prevent child sexual abuse.)

The reality is that some adults do not know the correct names for all the parts of the genitals. I once appeared on *The Today Show*, to discuss how parents can talk to small children about sexuality issues. I had shared my advice about the body-part game "this is your nose, this is your stomach..." After the taping, one of the cameramen asked if he could see me. "What was that word you used, besides the penis?" he asked worriedly. He had never heard the word "vulva" spoken aloud.

Many parents use the word "vagina" to refer to the external

female genitals. This is not correct. The vagina is the passageway from the external genitals to the uterus. It is not visible without a medical device called a speculum, which is used by gynecologists when they do pelvic exams on women.

"Vulva" is the correct word for the external female genitals. It includes the entire pelvic area: the inner and outer lips, the clitoris, and the vaginal and urethral openings. This distinction becomes important as young women reach puberty and adulthood; it is a good idea to start them out with the correct terms now so you do not have to correct misinformation later on.

Learning the correct names of the parts of the body will help give your child an ease with their body. A sexually healthy child (and indeed, a sexually healthy adult) feels comfortable with and appreciates their body. Recognizing that all parts of the body are equally special helps develop that sense of appreciation.

You may face some resistance from others about using these terms. One parent told me, "But they sound so dirty!" Another said that her own mother made fun of her. And your toddler may even have to correct other adults; I remember Alyssa saying to me at about age two, "Mommy, how come Bambi [her child-care provider] tells me to wipe my vagina? It's my vulva."

Some parents have asked me whether it is also important to use the more scientific terms *urination* and *defecation* during toilet training. Although I would like to say yes to be consistent with my views on teaching the names of the genitals, the reality is that four-syllable words are beyond the reach of most toddlers. (Vulva and penis have two syllables, and toddlers can pronounce them, honest.) In this case, I would suggest that you have a discussion with your partner and agree to the words you will use about the toilet. Just don't use words that convey negative feelings about the elimination of body wastes. We mostly asked, "Do you need to use the bathroom?" You can introduce the more complex, "real" terms when they are in elementary school.

Genital Exploration

Babies begin to explore their genitals during diaper changes. This usually happens at about seven to ten months, a little after they discover that they have fingers and toes. They experience for themselves that it feels good to touch all the parts of their bodies.

Many parents have told me that they are uncomfortable when they see their babies touch their genitals. They wonder what, if anything, they should do. They wonder if their child is masturbating or if the child is becoming obsessed with this behavior. This type of genital exploration is not the same thing as adult masturbation: It is generally not purposeful and it is not directed at orgasm. It is about exploring and learning more about the body.

Let's think about how *you* want to handle these situations.

First, ask yourself (and your partner), "What messages do we want our child to begin to learn about touching their genitals?" What are your family's values and feelings about this behavior?

Some parents want to give a message that touching and pleasuring yourself in this way is unacceptable: They move the child's hand, say "no" in a stern voice, and then go on with the diapering. (I have also heard of parents who slap their children's hands or put ice on their son's penises, behavior that moves beyond a negative message about masturbation to one that is sure to instill feelings of shame in the small child.)

Some parents choose to be silent about the behavior: They

move the child's hand away, say nothing, and continue on with the diapering. Other parents want their infants to delight equally in all the parts of their body, and they leave their children with plenty of undiapered time to allow for this exploration.

Which feels comfortable to you?

It is important to understand a few facts as you think through how you want to handle these situations. As children begin to discover their own bodies, it is as natural for them to touch and discover their genitals as it is for them to touch and discover their fingers, toes, and stomach. Understand that infants are not "masturbating" in the adult sense of the word; they are simply learning that it feels good to touch all the parts of their bodies. And know that there is no research to connect this infant exploration and later child or adult masturbation or other sexual behaviors. (In the next chapter, I will talk more about how you might want to respond to this behavior in toddlers and preschoolers as it does become more purposeful.)

What did we do in our home? We wanted our children to feel good about all the parts of their bodies, but we weren't quite comfortable with leaving them with their diapers off for extensive exploration time. We walked a middle ground: I'd smile and say something like "yes, it feels good to touch your body all over," and then I'd continue diapering.

Some parents also feel uncomfortable when their sons get erections during diaper changes. First some facts: Baby boys get erections every ninety minutes or so. Their erections are not a response to erotic stimulation: They are natural responses to touch, friction, or the need to urinate. Know that you aren't doing anything to cause the erection and feel reassured that you don't need to do anything to respond to it. Just keep on with the task at hand.

While we are on diaper changes, try not to give your child negative feelings about the elimination of body wastes. Making a face when you change dirty diapers or looking disgusted may give your child a message that something is wrong with their body. Try to substitute "Won't a clean diaper feel nice!" for "Whoa, isn't that a smelly diaper." Being calm and natural helps them learn that everyone urinates and defecates; it is just part of life.

Thinking through Gender Roles

During the first three years of life, children are learning the difference between boys and girls, and beginning to identify themselves as male or female. By the age of three, they know that they will grow up to be a man or a woman, and they are absorbing a remarkable number of gender stereotypes.

This shouldn't be too surprising. As I discussed earlier in this chapter, many families rigidly dress their children in gender-based colors, give them gender-based toys, and talk and handle them differently. The television and videos they watch also teach them about male and female roles. Think about it: Even on PBS's acclaimed *Sesame Street*, few of the muppets are female!

Children learn to divide the world into "male" and "female" before they reach the age of two. Depending on the behaviors they see in their homes and preschools, they identify certain behaviors as male or female. For example, in some studies, children as young as two have told researchers that girls like to play with dolls while boys like to play with trucks. They say that girls like to cry and boys like to hit. They divide adult jobs into male jobs and female jobs. They even divide colors into gender: Pink and purple stuffed animals are deemed girls; black and brown stuffed animals are deemed boys. And they may even change their behaviors to fit these gender stereotypes: For example, girls and boys in the first eighteen months have basically identical levels of aggressive behavior, but by the age of two or three, girls act less aggressive than boys. Both boys and girls seem to have learned that aggression is acceptable behavior for boys but not for girls. And there is some evidence that there are indeed differences in male and female brains and testosterone levels that account for some of the differences in behaviors by children of different sexes.

It is almost a cliché to say that children learn from what they observe around them. But research clearly shows that children from traditional homes are much more likely to learn these gender stereotypes. Parents often provide very different types of toys to girl and boy infants and toddlers. Girls get dolls, stuffed animals, kitchen appliances, jewelry, and dress-up clothes. Boys are given cars, trucks, balls, and sports equipment. I once did a study of toy catalogs: Almost every catalog rigidly divided the pages into toys that were only played with by boys and toys that were only played with by girls. Girls were sometimes pictured as doing "boys'" activities like playing ball or building science projects, but no boys were pictured with dolls, kitchens, or other "playing house" toys. In every catalog, the erector set was accompanied by a picture of a boy. Many psychologists believe that these toy selections actually affect skills in later life: Girls learn more about nurturing, and boys have more of an opportunity to develop reasoning and spatial-relation skills. (I often wonder if I would have a better sense of direction if I had been given building-set toys as a child!)

Interestingly, even children in less traditional homes still divide the world by gender based on the models they see. For example, in homes where the dad is the primary caregiver, small children will say that men grow up to be dads and moms go to work. I remember one day when two-year-old Gregory told me that he wished he could be a doctor, but he knew he couldn't. I asked him why. He, thinking of his own woman pediatrician, said, "Mom, only ladies can be doctors!"

It is important for you to think through these issues. Here are some questions to think about:

- What do you want to teach your child about what it is to be a woman?
- What do you want to teach your child about what it is to be a man?
- Do you want to limit your child's options as they grow up?
- Do you believe that boys and girls should have equal opportunities?
- Do you want your son to have the chance to role-play nurturing behavior? Are you comfortable giving him dolls to play with?
- Will your daughter drive a car when she grows up? Will she need to know how to fix things? Are you comfortable giving her fix-it toys and trucks to play with?
- How will you handle comments from other adults about a child's nontraditional activities?

Special Issue

Your Child-Care Provider

According to the U.S. Census Bureau, nearly six in ten families with children under six now have two parents who work outside of the home. During the day, our children are being cared for in day-care centers, by other mothers in their homes (family-based child-care programs), or by baby-sitters and nannies. Talking with your child-care providers about how you expect them to handle sexual issues is critical to assuring that your children's sexual education reflects your family values.

I learned this one the hard way. I was diapering two-and-a-half–year-old Gregory when he proudly pointed at his buttocks, giggled, and said, "ooh, you touched my heinie." I could scarcely believe my ears. Here I am, Ms. Use the Right Terms herself, and here is my son, using this funny, not altogether pleasant-sounding euphemism for his buttocks.

I discovered that Sharon, his nineteen-year-old nanny, who was doing as many diaper changes and baths as we were, was using the words her mother had used with her. I had frankly not thought to talk with her about our philosophy of raising our children to be sexually healthy adults, even though I had extensively shared and asked her to be consistent with our philosophy on such issues as discipline, toilet training, and healthy eating.

If you work outside of the home, it is important that you talk about sexuality issues with your child's primary care provider. You want to ensure that your child is getting consistent messages about such important issues as their body, genital touching, and gender roles. You want to stress that your caregiver should never, ever hit or shake your child; you also do not want them to be yelling, "Ooh, gross!" during diaper changes or telling your child that they are nasty if they touch their genitals.

You may want to ask them to read this book, or you might want to make a list of values that you want your provider to share with your child. Explain that these are your family values, and that they may be different from what she or he believes or learned in their own home. Tell providers that part of their job is to support you in giving your family's values to your child.

You might want to duplicate and complete this worksheet with your partner and share it with your baby-sitter, child-care provider, or nanny:

Sexual Issues and Our Child

We want you to use these words when you are diapering our child:

Penis or _____ Vulva or _____ Buttocks or _____

We want you to give our child the following messages about his/her body:

_____

If our child touches his/her genitals during a bath or diaper change, we want you to:

_____

Values Exercise for Chapter 3

Your four-year-old son says, "I want to be a ballerina when I grow up." You answer...
- ❏ a) "I'd like you to be a doctor or a lawyer."
- ❏ b) "Only girls can be ballerinas."
- ❏ c) "You'll change your mind many times."
- ❏ d) "Great! Let's look into dance lessons."

Your three-year-old daughter complains, "I wish I had a penis." You answer...
- ❏ a) "Don't be silly. Only boys have penises."
- ❏ b) "Where did you hear the word penis?"
- ❏ c) "Boys have penises; girls have vulvas."
- ❏ d) "Are you wondering about the difference between boys and girls?"

You walk into your four-year-old child's room and find him and the preschool girl next door with their clothes off. You...
- ❏ a) yell at them, "Get dressed immediately, and Dana, go home right now."
- ❏ b) tell them nicely to get dressed and distract them with another activity.
- ❏ c) ask them nicely to get dressed and then show them some books with pictures of different children's bodies.
- ❏ d) close the door quietly and let them continue with their play.

Your son or daughter is touching their genitals in the supermarket. You...
- ❏ a) hit their hand.
- ❏ b) whisper to them to stop.
- ❏ c) ignore their behavior.
- ❏ d) say, "I know that feels good, but please only do it when you are alone in your room."

Chapter 3
The Preschool Years
Ages 2 to 5

Preschoolers are very sexual people! During the years from ages three to five, your child will begin to develop a strong sense of whether they are a boy or a girl and what that means to them and the people around them. They are finishing with toilet training, and if they are in a school or a day-care center, they have probably observed other children and learned for themselves that boys and girls have different body parts. They are very curious about their own and other people's bodies. This is the age when your child is likely to first ask, "Where did I come from?" and it may be the time when your child explores sex play with a neighbor child or sibling.

During the preschool years, you will have many more opportunities to provide a beginning sexuality education to your children. With my own children, I have been surprised by how often "teachable moments" arise at these ages. Preschoolers are very curious about *everything*: They want to know why the sky is blue, why it snows, why that man is in a wheelchair, and why girls don't have penises. Anticipating some of these potential teachable moments will allow you to give *your* messages about sexually in a calm, relaxed way.

Teaching the Parts of the Body

I am always a bit surprised by how many adults either do not know or do not use the correct names of the body. Although all parents describe the elbow or the nose to their preschoolers in pretty much the same way, when it comes to the genitals, euphemisms abound. Parents say "ding dong," "weiner," "wee," and countless other words to talk to their little boys about their penises. Girls' genitals are less likely to have such cute names; many parents resort to "down there" or "your privates." Even books and preschools may talk about "private parts" rather than correctly labeling the genitals.

In the previous chapter, I talked about why I think it's important to label all the parts of the body equally. (If you skipped to this chapter because your child is now a preschooler, you may want to go back and read pages 34 to 37.)

However, new opportunities for teaching your children about their bodies arise during the preschool years. You can begin to talk to your children more explicitly about their bodies, as well as addressing their natural curiosity about the physical differences between boys and girls.

There are many opportunities to teach your preschooler about his or her body. Teachable moments for labeling all of the body parts include bath time, changing clothes, and having your preschooler help you diaper the new baby: "You are a boy and you have a penis; Mary is a girl, she has a vulva." If your child seems interested, you can continue, "All boys and men have penises; all girls and women have vulvas."

As your preschooler gets a little older, you can introduce

additional body parts. You might say to your son, "Those sacs between your legs are the scrotum; inside them are special parts called the testicles." If you see your three-year-old son with an erection, you can give basic information that lets him know this is perfectly natural: "Sometimes your penis is soft, sometimes it's hard." You might tell your four-year-old daughter during a bath: "That opening between your legs is called a vagina. That little button on the top of the lips is your clitoris."

Try to be calm and matter of fact. You want to try to convey a message that all the parts of the body are good and special, and that all the parts of the body have their own names.

Children's books and anatomically correct dolls provide another way to introduce all the parts of the body, including the genitals. You may want to get a copy of one of the books listed in the appendix. And anatomically correct dolls can help you teach that in real life, everyone's body has genitals.

It actually may be hard to find anatomically correct dolls. On a recent shopping trip, I was unable to find them at Toys "R" Us or FAO Schwartz. You may have more luck in specialty toy stores. These dolls also tend to be expensive: The cheapest one I found was $29, and some were selling for as high as $80. As an alternative, you may want to contact an organization called Teach-A-Bodies, which makes anatomically correct dolls (see the appendix).

Teaching about body parts is an ongoing process. Your children will not learn it all at once. One of my colleagues told me a story about his then two-year-old daughter. During a bath, she asked him, "Daddy, what's inside my vulva?" He answered, "We've talked about that before. What do you remember?" She answered, "Well, you told me there are some critters in there!" He told me that he was pretty sure that is how she remembered the word "clitoris."

So repetition is important. You, your partner, and your child-care provider need to use the correct names of the parts of the body when you help your child change their underwear, observe them touching themselves, or teach them basic hygiene techniques (e.g., "when you wipe your vulva, wipe from front to back"). You don't need to force this or artificially create situations; they will arise naturally. Just try not to slip back into the euphemisms that your parents probably used with you.

You may need to help your child understand why other children use such funny-sounding names for the genitals. In preschool or kindergarten, they may hear other children using these euphemisms. They may even hear adults using the wrong names. For example, Alyssa wanted to know why her baby-sitter used the word "vulva" when she was wiping her vagina.

One of my colleagues told me this story: Her son Bruce had only heard the word "penis" used in their home. He came home from kindergarten one day and asked, "Mom, what's that man's name that people use for penis?" My colleague swallowed hard, wondering what her son had heard. He continued, "You know that man's name . . . is it Bob?" She thought a minute, concerned that he had perhaps overheard her in her bedroom using an endearment for her husband's penis. Then she relaxed. "Oh," she realized, "you mean Dick."

Messages for Preschoolers about the Body
- Every part of the body has a name and a purpose.
- Boys' and girls' bodies have most of the same parts but a few that are different.
- Boys have a penis and a scrotum.
- Girls have a vulva, a vagina, and a clitoris.

Bath Time

Bath time can be a wonderful teachable moment for talking about the body.

Many parents have asked me if it is all right to bathe two children of the opposite sex together, and if so, at what age this should stop. Other parents have enjoyed baths or showers with their toddlers; they want to know if they should stop now that their child is at preschool age. There are no hard-and-fast rules here; like so many of these issues, it depends on your family's values and the nonverbal feedback you get from your children.

You need to ask yourself (and your partner) about your family values about nudity. Are you only nude in the shower? Do you walk around without your clothes on frequently? Are you comfortable bathing or changing clothes with your children in the room? What messages do you want to give your children about nudity and their bodies? Is it important for your children to feel comfortable around each other without clothes on so that you can easily change and bathe them together, or conversely, is it important to you that they only do so privately? Remember to take off your adult lenses as you think about the bath question. Your children will not become sexually aroused bathing together. Remind yourself that these are not two nude adults in a hot tub. But know that it is natural for them to be curious about each other's bodies and the differences they observe.

Pay attention to your children's cues. If your children seem relaxed and comfortable bathing together, there is no reason to separate children of preschool age during baths. It can actually be an easy way to introduce that boys' and girls' bodies are mostly the same, but that only boys have penises and only girls have vulvas. However, if your children start to seem uncomfortable or giggly, they are probably giving you a message that coed bathing should be coming to an end.

The same principle applies to parents' bathing and shower-ing with their children. Shared baths can be a lovely time for talking and relaxing together. But sometime during your child's preschool years, he or she may become uncomfortable with your grown-up nude body. Or they may want to touch you in a way that feels inappropriate or uncomfortable to you. It is nat-ural for a child to be curious about a mother's breasts or their father's penis, or why adults' bodies have pubic hair and theirs doesn't. They have a natural curiosity about how adult bodies and children's bodies differ. You need to decide if you feel com-fortable with handling these questions about these differences and if your children seem comfortable with your adult nude body. By being aware of your child's often unspoken messages, you will know when you've both outgrown this practice.

At a recent parent workshop that I led, several mothers were worried that their preschool child had reached out to touch their adult breasts or genitals. One mother was particu-larly concerned that her three-year-old son kept trying to put his hand down the front of her pants. I asked her, "Why do you think he is doing that?" She suggested that it might be because he was curious or it might be because he liked to provoke the kind of intense attention it aroused. I then asked her if she was still washing her son's genitals. She said that she was still wash-ing his penis during baths and helped him wipe himself after he used the toilet. I told her that it was possible that all or any of these might be contributing to his inappropriate touching and that she might try to address all of these. She could teach her son that no one should touch anyone else's genitals and begin to teach him self-care of his penis. She could share some age-appropriate books with him about bodies, privacy, and touch-ing. And, she could tell him that this behavior made her uncomfortable and that she would like it to stop.

Three- and four-year-old children can also learn that it is important for them to keep their own genitals clean and healthy. When you are bathing or showering your child, give them the washcloth and have them wash their own vulva or penis. They are old enough now to begin this type of self-care. And while you are at it, be sure to teach them to wash from front to back to avoid spreading germs.

Having children wash their own genitals provides another teachable moment as well. You can use this time to introduce the concept that no one but the child should touch his or her genitals: "You are old enough now to wash your own penis/vulva. Your body belongs to you. No one should touch your genitals except a doctor or nurse or mom or dad for health reasons." And you can start teaching them that these parts of the body are private.

"I Want to Be Alone!"

Three- and four-year-olds are not too young to be introduced to the concept of privacy. By the time your child is a preschooler, you may be desperate to be able to use the bathroom alone—I know I was. You may be increasingly uncomfortable about your growing child seeing you in varying states of undress. You probably want to give yourself and your partner some time alone together. You can also now introduce the idea that some behaviors, like touching your own genitals, only happen in private.

Different families have different values about privacy. You will want to ask yourself and your partner:

- How do we feel about our children's watching us dress? undress?

- What will we do if a child walks in when we are making love?
- Is using the bathroom a solitary behavior in our home or can someone come in when we are using the toilet?
- How much privacy are we willing to give to our preschool children? Can they play in their rooms with the door closed?

And know that even as you set the parameters in your own home, your children may be faced with situations with different rules. For example, we had taught three-year-old Alyssa that people use the toilet alone with the door shut. When she entered a new preschool, the teacher reported that she was having a difficult time getting Alyssa to use the bathroom. We discovered that the school's policy was to line all of the children up, leave the door open, and have two children use toilets side by side. No wonder she was resistant when this was so different than what she had learned in our home. We tried to convince the school to let her shut the door; they explained that their policy was designed to assure that the adults who worked at the school could not be accused of inappropriately touching children.

So, what did we do? We used it as an opportunity to teach Alyssa that people have differing values about personal issues. We told her, "Your school has a policy that says they have to leave the door open. Because you need to use the toilet during the day, you will have to follow that at school. But in our home, we use the bathroom with the door closed." And in a few weeks, she got used to both.

This introduction to the issue of privacy also helps in responding to your child's inappropriate touching of their genitals in public.

Genital Touching

Three- and four-year-olds delight in their own bodies. And just as they learn it feels good when their bodies run, jump, and cuddle, they are also discovering that touching their own genitals feels good. Most professionals believe that preschool children touch their genitals as a natural part of child development. In a recent study of mothers of two- to five-year-olds, about 25 percent of moms reported that their son and 15 percent reported that their daughter touched their own genitals in public. Sixteen percent of both the preschool boys and girls had been observed touching their own genitals.

Many adults see this behavior and think of it as "masturbation." I am not going to use that term as I talk about preschoolers, because it tends to conjure up pictures of adult and adolescent self-pleasuring aimed at achieving orgasm. Instead, let's re-label this behavior "genital touching." (There are no real statistics on the transition from genital touching to masturbation. Many sexologists believe that full-scale orgasm isn't possible before puberty, although, conversely, many adults report that they remember having their first orgasm before their teenage years.)

Toddlers and preschoolers touch their genitals in a much less purposeful way than older children and adolescents, and many do so without embarrassment or anxiety. In fact, it is not unusual in a preschool to have several of the boys touching their penises unconsciously throughout the day. Little girls find it pleasurable to touch their vulva and clitoris as well: I often get calls from parents who are worried because their three- or four-year-old daughter seems to be rubbing her vulva against the couch or pillows in the living room.

Many children appear to not even be aware that they are touching their genitals in this way. Some children do discover that this activity helps them calm down: Many children touch their genitals right before naps or bedtimes to help them fall asleep. (Just like adults, some children never touch their genitals at all. It is normal for a preschooler to touch their genitals, and it is normal for them not to. Please do not worry that your child is not developing normally if you don't see them touching their genitals!)

Parents respond to a preschool child's touching his or her genitals in a variety of ways. Some parents are concerned that their child, in the words of one parent in one of my workshops, is becoming "obsessed with sex." Others are faintly amused that their child has discovered this pleasure on their own. Still others are annoyed: "I don't mind that she does it, but why does it always have to be around her grandmother?" or "How can I get my son to stop holding onto his penis?"

So, how do you respond to this behavior? First, you need to think about your family's values about masturbation. Are you comfortable with genital touching as a form of self-pleasuring? Or do you believe that it is a behavior to be discouraged? Which messages do you want to communicate to your children?

Regardless of your family's attitudes, all children need to learn that touching their genitals is a private behavior, just like using the bathroom is a private behavior. Preschoolers can understand that other people will be upset if they see them touching their genitals in public, and that this type of behavior should be reserved for when they are alone in their rooms.

Many professionals recommend that if your child is touching their genitals in a private place and you come upon them, you simply ignore the behavior. You might just leave the room quietly or say, "I'll come back in a little while." This is one time where I do not recommend commenting on the behavior;

most of us would not be comfortable saying, "I see you are masturbating; good for you! I'll leave you alone now."

However, many preschool-age children touch their genitals in public. Many parents of preschool-age boys have complained to me that they can't get their son to stop touching his penis at home or school. If your child is touching his or her genitals in the supermarket or in the living room, you can gently try to stop the behavior. First, quietly acknowledge it in case they are not aware they are doing it: "Steven, you are holding your penis again." Then, remind your child that this is a private behavior. You might say something like "I know it feels good to touch your penis, Chris, but I want you to stop. Touching your penis should only be done in private. What's a private place in our house?" or "I'm uncomfortable, Susie, when you rub your vulva against the couch in the living room. Touching your vulva should only be done in private. What's a private place in our house?"

You will probably have to repeat this lesson many times before your child can distinguish between public and private places. After all, we don't expect our children to remember to look both ways when they cross the street after only telling them once!

Some children seem to touch their genitals continuously throughout the day, often to the exclusion of other activities. Some children do this as a way to calm themselves during an emotional time, such as when there is a new baby or a divorce, just the way other children will begin compulsive thumb-sucking or hair-twirling as a way to deal with such stresses. However, in other children, it may be a sign that they have been sexually abused. If this is even a remote possibility, please skip ahead to the end of this chapter's Special Issues section on sexual abuse on page 82.

> **Messages for Preschoolers about Genital Touching**
> - It feels good to touch parts of your body, including your penis or vulva.
> - This type of touching should only be done in private.
> - Private places are places where you are by yourself. In our home, a private place is _____ (fill in).

"We're Just Playing Doctor, Mom"

The curiosity that is part of genital touching also sometimes leads to games with other children. Many adults remember playing doctor or house. Sometimes these games turned sexual: I remember playing "fraternity" with my sister and our two best girlfriends when she and I were four and six. Two of us would be the girls, two of us would pretend to be the boys. We would pretend to go on "dates" and often we would kiss each other or undress in what we thought were provocative ways. (To this day, I hesitate to write this, as I think about my own parents' finding out about our very clandestine behavior!)

My friends and colleagues report similar game playing. "I'll show you mine, if you show me yours" is not just a game in the movies. Playing doctor, including undressing each other and examining each other's genitals, is quite common. It happens between boys and girls, boys and boys, and girls and girls, and it tells one nothing about future adult sexual orientation. As many as half of adults remember engaging in childhood sexual play. It is interesting to note that throughout the years, playing doctor seems to be the usual scenario for most of this sex play. Without an adult sexual context, children know that doctors look at patients' genitals, which is further evidence that this behavior usually indicates curiosity, not a desire for erotic fulfillment. Most child-development experts see this type of sex play as expected and natural childhood sexual curiosity. If the children are of the same age and stage of development, experts do not believe that it is harmful. And studies of adults show that engaging in childhood sexual play doesn't seem to have any impact, either positive or negative, on adult sexuality.

What's going on here? Preschool-age children are curious about their bodies and about other people's bodies. They may especially be curious about the bodies of the other gender. At this age, they are also mimicking adult behavior as they play house, fire fighter, and doctor. They are trying on roles and behaviors. Putting curiosity and role-playing together often leads to what some experts have labeled "childhood sex play." This curiosity leads to touching, and children may discover that this type of touching feels good.

In fact, some sexologists even believe that early childhood sex play teaches children some important skills. They point to studies of monkeys: In monkeys, early (pre-adult) sex play lays the foundation for successful male-female reproduction in adulthood. Monkeys who are raised in isolation, and do not have this opportunity, never copulate or reproduce, even when paired with an experienced mate. And although I certainly do not believe that children *need* this type of behavior to become sexually healthy adults, I agree with those professionals who say that most of it is harmless.

So, what do you do if you walk in and your child and the neighbor's child are undressed together, or maybe more disturbing, touching each other's genitals? First, take a breath; try to calm down. Try hard, really hard, not to look at this through adult lenses. Your three-year-old is not having sex with the boy next door! She is most likely curious about how his body is different than hers and vice versa. Start by saying, in as composed a voice as you can muster, "Tell me what game you are playing."

Let's think together about which messages you want to convey. Many adults can remember a situation when they were discovered playing one of these childhood sexual games. They can still remember the shame and guilt they felt as their parents screamed at them to stop. Some sex therapists report that their adult clients still remember and are still influenced by the messages of shame and guilt that they were given as preschoolers when they were discovered at a childhood "show me" game.

So, what do you do? I suggest calmly telling the children to get dressed and to come into the living room. I would acknowledge their curiosity, but suggest that there are other, better ways to learn about boy and girl bodies. You might want to share one of the books in the appendix with the children. You might also want to reduce their time alone together in rooms with closed doors.

And, yes, I think you need to tell your neighbor. If the situation were reversed, wouldn't you want to know? Don't blame the other parent for the behavior, but don't let them blame you

---

A Caution

There is a big difference between harmless child sexual play and play that is exploitative or abusive. Children do not naturally engage in painful sexual behavior, oral-genital contact, or simulated or real intercourse or penetration with fingers or objects with another child. These activities could indicate exposure to inappropriate television or videos. In the worst-case scenario, they may indicate that a child has been sexually abused. Nor do most children engage in curious normal sexual play with children more than a few years older than they are. If your preschool-age (or elementary school–age) child exhibits any of these behaviors, you should contact your pediatrician for an evaluation.

either. Explain what you saw and tell them you don't know who started it. Share these pages from this book and assure them that this is expected behavior. Try to agree to some "rules" about the children's playing together: public spaces, fully clothed, no locked doors.

Some parents have called me especially concerned when they find that preschool sex play is going on between siblings. Again, this is a good time to take off your adult glasses: This is not incestuous behavior. It is the same type of childhood curiosity as play with neighbors; it usually happens because siblings close in age may be more available for this type of play than other neighborhood children. You can deal with it in the same way you would with a neighbor's child, although you might want to cut down on unsupervised times such as shared baths and naps in the same bed.

And just as you may walk in and discover your child and another child engaged in sex play, the reverse is possible too. Sooner or later, almost all parents have the experience of their child's surprising them in the middle of adult lovemaking!

Messages for Preschoolers about Sexual Play
- Children often kiss, hug, and touch one another in ways that feel good.
- Children are often curious about each other's bodies.
- It is not okay to hug or kiss someone if they don't want you to.
- Your body belongs to you.
- You have the right to decide if another child may touch your body during a game or any other time.
- When playing at home with other children or outside, children keep their clothes on.

Preschool and Early-Elementary Sex Play:
Harmless or Problematic?
Here's a quick way to assess whether childhood sex play is
likely to be harmless or whether you should be concerned:

	EXPECTED	MAY BE PROBLEMATIC
Ages of children	Similar	More than three years apart
Children seem	Giggly, curious, happy	Aggressive, angry, fearful, withdrawn
Activities	Undressing; playing doctor or "You show me yours, I'll show you mine"	Oral, anal, or vaginal intercourse; penetration with fingers or objects
After discussion with parents	Behavior stops	Behavior continues

Friendships and Feelings

Of course, sex play is only a tiny part of preschool friendships and play. Preschoolers are beginning to develop
friendships with other boys and girls. Although babies enjoy it
when older children play with them, and toddlers like to play
sitting side by side with another child, it is around the age of
three that children begin to develop the capacity for selecting
their own friends. Preschoolers begin to choose which children
they like to play with and which children they don't. Preschool
can also be an important time to help them begin to develop
early friendship skills.

At around the age of three, many children begin to group
themselves by gender in play situations when they have a
choice of friends: Girls play together and boys play together.

Boys and girls often will choose different activities, and they often have different styles of play. Boys tend to play in larger groups and their play can be rougher. Girls tend to prefer to play with one or two other girls. As a result, your child is likely to prefer to play with children of the same gender. However, with some encouragement, most children will play easily some of the time with children of both genders. This is important for laying the foundation for friendships with people of both genders when they are adolescents and adults.

Parents can encourage boys and girls to continue to interact together in pleasant ways. Invite a child of the other gender over for a play date. Encourage groups of children at the park or playground to work on a project together or play a large group game. Beginning friendships with people of both genders helps set the stage for healthy, respectful relationships between the genders in adulthood.

If your child is in day care or at a preschool, talk with the teacher about how children are encouraged to form friendships. Ask if there are opportunities for boys and girls to play together and if all the parts of the classroom are open to both boys and girls. Suggest that at least some of the time the class be split up by characteristics other than gender: For example, groups can be divided by the colors of their shirts (the red and the blue shirts form one line, the green and the brown the other) or their types of shoes (all the sneakers will go to the doll corner, all the boots will go to the blocks).

These beginning friendships are also an opportunity to talk to your child about feelings. It is not at all uncommon at this age for children to have their feelings hurt.

I know I will never forget the first time another child hurt Alyssa's feelings. We were having supper at a friend's house who had a daughter who was about two years older than Alyssa.

They had always played nicely together, but this night, a third child who was even older was visiting as well. The two older children went to Katie's room, shut the door, and left Alyssa sitting outside. I came upon her, softly crying outside the bedroom door. It was her first experience with rejection, and I was flooded with my own memories of every time that I had ever been rejected as a child.

I even had to fight off my own feelings of being angry at Katie and her mother. I was struck by the thought that this might be the first time Alyssa was rejected, but it certainly wasn't going to be her last.

I remembered to use it as a teachable moment.

Me: How do you feel, honey?
Alyssa: Mommy, I'm so mad I want to yell at them.
Me: I bet you feel sad too.

Talking about and labeling feelings is important for preschoolers. Three- and four-year-olds can begin to label their feelings and begin to identify the feelings of others. They are feeling the emotions that mean "happy," "sad," "angry," and "mad," and they can learn to talk about what causes them to feel this way. They also can identify nonverbal emotional cues in others; it was not unusual for four-year-old Gregory to ask me, "Mom, how come you look sad today?" after I came home from a particularly stressful day. Preschoolers are beginning to learn empathy, an important quality in adolescent and adults. Children at this age can even sometimes predict which situations will evoke a certain kind of emotion: "Are you angry that I spilled my milk on the floor?"

Parents can help their children label and express their feelings. One thing you can do is to try to be sensitive to your child's emotions and help them label their feelings. You could say something like,

> I think you're feeling _____ right now; is that right?
>
> You seem really _____ now.
>
> It makes you really _____ when you _____ .

You can also label the feelings of characters in books and videos or ask your child to guess what the character is probably feeling in the situation: "How do you think Belle feels when she sees the Beast for the first time?" "How do you think the Lion Cub feels when he finds out his father has died?"

It is also a good idea to periodically label and explain your own feelings as you interact with your child. "I'm feeling a little sad today because your daddy is out of town on business" or "I'm

---

Messages for Preschoolers about Friendship and Feelings
- Friends have fun together.
- Friends help each other.
- Boys and girls can be friends with each other.
- Some children have a lot of friends, and some have a few.
- It can hurt someone's feelings if you tell them that they cannot play with you.
- People have many feelings: They can be happy, sad, angry, excited, lonely, hurt, confused, or frustrated.
- Making other people happy can make you feel happy, too.
- It is okay to feel angry; it is not okay to hurt someone.
- Everyone feels scared sometimes.
- Words can help us describe and share our feelings.
- It is good to tell people about your feelings.

---

sorry I yelled at you; I was angry because you weren't listening." It never hurts if you apologize to your child. It lets them know that nobody is perfect and it models how to say "I'm sorry," something that three-year-olds don't see any reason to do.

These early lessons about expressing one's feelings set the foundation for the future. Children who can identify their own feelings and anticipate the feelings of others become teens and adults who know how to listen and respect others and demonstrate empathy and compassion.

"When I Grow Up, I'm Going to Marry Mommy"

Preschoolers also have intense feelings for their parents. When he was three, Gregory proudly announced to our family that he was going to grow up and marry me. You probably remember reading in college introductory psychology classes about the Oedipus complex and the Electra complex. Sigmund Freud felt that three- to five-year-old boys develop fantasies of possessing their mothers sexually and become jealous of their fathers. (He named it the Oedipus complex after the Greek tragedy in which Oedipus unknowingly kills his father and then marries his mother.) According to Freud's theory, this period is accompanied by castration anxiety: Boys fear their fathers will retaliate against their son's interest in their mothers by cutting off their penises. Freud said girls develop an Electra complex: They develop "penis envy" and want to take their fathers away from their mothers. Girls, he said, reject their mothers because they blame them for their lack of a penis. (Freud named this the Electra complex after the Greek tragedy in which a princess helps kill her mother.)

Most modern psychologists debunk these ideas. Feminist psychologists critique Freud's views as extremely sexist, and anthropologists point out that these "complexes" are not found in all cultures. Still, it is not uncommon for preschool children to prefer the parent of the other sex at this time.

Preschool children love their parents intensely and may be confused about whether all love is expressed romantically. Many children try to kiss their parents romantically. Some even try to imitate adults they've seen in person or on TV or in movies by putting their tongue in their parent's mouth. It is important to explain to children that this type of kissing is not appropriate for children together or children with adults: "I don't feel comfortable with you kissing me this way. This is a way that adults kiss. I'd like you to kiss me on the cheek."

Preschool children sometimes may even seem jealous of their parents' showing affection to each other. In our home, Gregory often interrupted my husband and me hugging with a request for a "family hug." In one study, 13 percent of mothers of two- to five-year-olds reported that their child gets upset when they see adults kiss.

It is important to reciprocate your child's love but not their romantic feelings. You might respond to your son's declaration like this: "I love you a lot. It's fun to think about who you will marry when you grow up. When you grow up, you will probably fall in love and marry a terrific woman."

This is also an opportunity to learn more about what your child is thinking. I had the following discussion with four-year-old Gregory:

Gregory: Mom, I wish I could marry you when I grow up.
Me: I love you a lot, Greg, but I'm already married to Daddy.
Gregory: But I couldn't marry you even if you weren't married to Daddy.
Me: That's right, honey.

Gregory: Right, because you go to work and I don't want to
   marry someone who goes to work!

Me: Why don't you want to marry a woman who works?

Gregory: Because I want her to play with me all the time!

Me: Are you feeling sad that I can't play with you all the
   time?

Obviously, there were many issues in this discussion. It is
important to let your child take the lead in these discussions
and to pay attention to their feelings. Really try to listen to
what they are saying.

These declarations can also be a teachable moment to talk
with your children about the roles of men and women and
about what the children want to be when they grow up. Adults
frequently ask children, "What do you want to be when you
grow up?" And we are often somewhat amused by their
answers: So far Gregory has wanted to be a garbage collector, a
comedian, and an ice skater. I think it's important to not only
ask children about their future career choices, but also to talk
with them about whether or not they want to be a parent when
they grow up. You can introduce this by asking, "And do you
want to work and be a daddy too?"

This can also a teachable moment to provide your child
with your values about love. People often forget to talk to their
children about love. We tell them we love them, but we assume
that they know what we mean. Love is actually a pretty
abstract concept, but even a preschool-age child understands it
as a feeling. One easy definition I know and like is "Love is the
happiness inside our heart."

Think about all the stories for preschoolers that deal with
love and marriage. Belle falls in love with the Beast; Cinderella
meets the handsome prince; the prince kisses Sleeping Beauty

and wakes her up. I have personally never been comfortable with all of the children's stories that end, "They got married and lived happily ever after." I think that too many adults spend their life looking for that perfect, problem-free marriage that they were promised as preschoolers. With my own children, I have substituted, "They got married, they were happy, and it was a lot of work!" or some more honest variation.

Your preschooler may also have their own fantasies about love and romance with their schoolmates. It is not uncommon for four- and five-year-olds to have crushes or boyfriends and girlfriends. They may even stage weddings. Four-year-old Gregory was very fickle; every few months or so, he told me he had a new wife at preschool.

One of my friends tells a story about coming upon her four-year-old daughter standing in front of an open refrigerator door, muttering "Danny, Danny, Danny." When my friend asked her what she was doing, Belinda answered, "I'm thinking happy thoughts." My friend asked her, "But why are you standing in front of an open refrigerator?" Belinda answered, "In *Peter Pan*, Peter said 'think happy thoughts and you can fly.' Well, I want the chocolate milk on the top shelf, and thinking about Danny makes me very happy!" At four, she already had a crush on a fellow classmate.

Some psychologists even believe that early-childhood crushes set the foundation for your lifelong "attraction template." An "attraction template" is a model for the type of person you find sexually attractive. Most people have a certain physical type that they find attractive. Think for a minute: Did you have a childhood sweetheart? Do you remember what he or she looked like? My first "boyfriend" in the third grade was Italian; according to this theory, he set the stage for my marrying my husband, Ralph Tartaglione!

Messages for Preschoolers on Love
- Love is the happiness for another person we feel in our heart.
- Children need to grow up with people who love them.
- People give and receive love.
- Love feels different when it is to your parents, other family members, pets, and friends.

"Is My Son Gay?"

Some parents of preschoolers worry that their child is never going to grow up and marry and have children. I often get calls from a worried parent of a four- or five-year-old boy or girl. They tell me, "My son likes to wear his mother's shoes," or "My daughter won't play with dolls," or "My son wants to take ballet lessons."

I gently ask them to tell me more about their concern. After a few minutes, they usually sheepishly ask, "Do you think my child is gay?" and then "What can I do to stop it?"

I usually first encourage these parents to relax. I give them some facts about sexual orientation: "You cannot cause your child to be gay any more than you can cause them to be heterosexual. And childhood behavior can be anything from a phase to a predictor of adult sexual and gender orientation."

Some definitions are in order here. "Sexual orientation" is usually defined as one's erotic, romantic, and affectional attraction to people of the same sex (homosexual), people of the other gender (heterosexual), or people of both sexes (bisexual). "Gender identity," on the other hand, is not about erotic attraction; rather it is about whether or not one understands oneself to be a male or female and whether or not one affirms for oneself the societal roles, values, and responsibilities of being a man or woman.

Most sexuality professionals prefer the term "sexual orientation" to "sexual preference" because we know that people do not choose whether they are homosexual, bisexual, or heterosexual. Most professionals believe that many factors determine sexual orientation. These include genetics, prenatal hormonal influences, sociocultural factors, psychological factors, or a combination of all of these. More and more science points to a genetic basis.

Sexual orientation has four basic components. It is defined by (1) who a person falls in love with, (2) who a person has sexual behaviors with, (3) who a person fantasizes about, and (4) how a person identifies themselves. So, if you fall in love with someone of the same sex, you share sexual behaviors with someone of the same sex, you fantasize only about people of the same sex, and you identify yourself as gay or lesbian, you are a homosexual. Or, if you do all those with an opposite-sex partner and identify yourself as a heterosexual, you are a heterosexual.

But it's not always that neat. Some people fall in love with people of both sexes, but only share behaviors with someone of the other sex. And some people love someone of the other sex, and only have sexual relations with that person, but occasionally fantasize about someone of the same sex. Any combination is possible.

Biological sex is not the same thing as either sexual orientation or gender orientation. It is determined by the chromosomes you are born with. People are usually born with XX chromosomes if they are female, or XY chromosomes if they are male. But some people are born as XXY, XXXY, XXXXY, or XXYY (these are boys with a disease called Klinefelter syndrome) or XO (these girls have a disease called Turner syndrome) or XYY (these boys may lag behind in physical development, do less well in school, and may be more likely to engage in violent behavior). Most babies are born with either a penis or a vulva and clitoris, but in about one in 1,500–2,000 cases, babies are born with parts of both sets of genitals (they are called intersexuals) or with external genitals that do not match their internal reproductive organs (they are called pseudo-hermaphrodites). In any of these cases, you will certainly have recognized these by the preschool years, and I hope you will have talked with a pediatrician who specializes in these syndromes.

Gender identity and gender orientation are different than sexual orientation and biological sex. Gender identity is defined by whether someone identifies as male or female, and it too is defined by many components.

By about age three, most children can identify whether they are a boy or a girl. By the time they are ages five to seven, most children have developed what psychologists have labeled "gender constancy." They know that they will always be a male or a female.

Although many children adapt their behaviors to conform to adults' gender expectations, some do not. In the words of the Children's National Medical Center, "children with gender-variant traits have strong and persistent behaviors that are typically associated with the other sex." In fact, this is much more common than many people think. Fourteen percent of mothers of boy preschoolers and 10 percent of mothers of girl preschoolers report that their child likes to dress up like the other gender, and 6 percent of boys and 8 percent of girls this age have said that they want to be the other gender.

By and large, girls are given considerably more leeway in behaving in nonstereotypical ways: A girl can hate wearing dresses, prefer blocks to dolls, and be physically active without her parents or care providers becoming alarmed. On the other hand, a four-year-old boy who prefers dolls, likes playing with girls more than boys, and wants to wear his mom's old dress may raise more concern. Just think about the words that people use: Girls are labeled, usually without fear, as "tomboys"; boys are labeled "sissies." Many dads, in particular, worry that their sons are not masculine enough.

Children of this age carefully watch the adults in their lives for gender cues. Your daughter may like to watch mom put on makeup; your son may like to watch dad urinate standing up. They also watch to see which behaviors women do and which men do. They divide the world into boy toys and girl toys and male occupations and female occupations. By the age of four, Gregory would no longer play with certain toys that he perceives are for girls; if the kids' meals at the fast-food restaurants only had girl toys left, he would choose to go without a prize rather than take the pink pony.

I believe that it is important to give your child a strong sense of what it is to be masculine or feminine without limiting their options. Most of this type of cross-gender behavior is role-playing; your child is trying on different behaviors. And many parents value giving their children a sense that they can be anything they want to be. In our home, I have taught my children that the only absolute differences between men and women is that men can go to the bathroom standing up and women have uteruses so that they carry the babies. Otherwise, we try to nurture their interests without regard to gender.

There are some children who will continue to develop a stronger identification with the other gender as they become adolescents and adults. About one in 100,000 men and one in 130,000 women are transsexual: For a variety of hormonal, prenatal, and possibly environmental factors, they do not feel that their actual physical sexual anatomy matches their gender. This feeling, if it persists into adolescence and adulthood, is only partially relieved by dressing or attempting to pass as a member of the other gender. (Transsexuals are not the same as transvestites, who are people who like to dress as the other gender for sexual arousal.) If your child repeatedly says that they hate their penis or their vulva or vagina, this might be a sign that you want to contact a health or mental-health professional. If your child consistently tells you that they want to grow up to be the other gender and seems acutely distressed when you tell them they won't, this might be a sign. In these cases, you will probably want to talk with your pediatrician about a referral to someone who specializes in these cases or what is known as a "gender dysphoria" program. There are people who believe that they are born transgendered and experience this difference very early in life: You may not be able to change them, but with help, they can live happy lives. PFLAG, the national organization of Parents, Families, and Friends of Lesbians and Gays, has an excellent pamphlet, "Our Trans Children," and can also offer you local referrals to experts. They are listed in the appendix.

The Children's National Medical Center in Washington, DC, sponsors the "Outreach Program for Gender-Variant Children and Their Families." They sponsor a listserv for parents with children whose behaviors are not typical of their gender. See their Web site, www.dcchildrens.com/gendervariance. In chapter 4, I'll talk more about "tomboys" and "sissy girls."

"Where Do I Come From?"

Almost every parent of a preschooler shares the same concern: How and what should I tell my child about reproduction? Some time between the ages of two and a half and five, your child will probably ask you this question, and it is good to be prepared with your response. Children at this age are often very curious about pregnancy and birth. According to psychologists, at around the age of four, children understand that babies do not just spontaneously appear and that something must have happened for this process to begin.

Psychologist Anne C. Bernstein did research with small children about how they understand reproduction. She labels the preschooler asking about reproduction a "geographer." The emphasis really is on the *where*: They want to know *where* the baby comes from and *where* it was before it was born.

Little children are known to be concrete thinkers and are very literal. One study found that children who are told that "a seed is planted in the mommy" actually picture plants growing inside their mothers. Being told that a baby grows in the mommy's stomach may frighten a child who associates stomachs with food and eating and may cause them to wonder why daddies, who also have stomachs, don't have babies.

Some parents have asked me why it is important to answer preschoolers' questions about where babies come from. After all, what could be wrong about saying, "You're too young to know this; I'll tell you later"? It's true: Avoiding this question may get you off the hook for the moment, but it also gives your child the message that you don't want to talk to them about sexuality issues. Answering this question simply now lays the foundation for future conversations, and it tells your child that you will teach them about this important subject.

Although "Where did I come from?" is a difficult question, many parents' most dreaded question is "But how does the baby get in there?" Be assured that preschool-age children are not looking for a detailed description of intercourse. In fact at this age, most are likely be disgusted by the very idea if it is presented in explicit detail. So, how do you answer the "how" question?

First, find out what your child already knows.

Remember that joke from the first chapter about the little boy who comes home and asks his dad, "Where did I come from?" The father launches into a fifteen-minute detailed explanation of anatomy and reproduction. The boy finally interrupts, "But, Danny says he's from Cleveland; where did I come from?" In other words, first make sure you know what your child is really asking! You could start by saying something like, "Where do you think you came from?" or "Do you mean where did we live when you were born?" or "Do you want to hear the story of how you were born?"

Second, remember the values you want to share with your child. For example, is it important that your child learn that intercourse belongs only in marriage or is only for adults? You can use this question to share those values. Do you have special family issues to consider: Was your child adopted? conceived through artificial insemination or in-vitro fertilization? delivered by c-section? Are you gay or lesbian parents? These types of situations could also affect your answer. (If you adopted your child or conceived him or her through assisted reproductive technologies, you may want to turn to the Special Issues section in the next chapter.)

Third, start off with very simple answers, and watch to see if your child is interested in continuing the discussion. Here is one way the conversation could possibly go:

Child: Where did I come from?
Parent: Do you mean where were you born or how do babies get started?
Child: I mean, babies.
Parent: What a good question. Babies grow in a special place inside a mom called the uterus.
Child: What's a uterus?

Parent: It's a special place inside a woman, right below her
  stomach. Only women have uteruses, so only women
  can have babies. But dads play a special role in
  helping a baby get started.

Now, the reality is that this type of conversation will satisfy
most three- to four-year-olds initially. But, don't be surprised if
they go away, think about it some more, and come back with
more questions. Indeed, be happy if they do!

Child: Daddy, I've been thinking, how does the baby get
  into the uterus?
Parent: A man and a woman are needed to start a baby.
  Inside the woman is a tiny egg cell; inside the man,
  are tiny sperm cells. When the egg cell and the sperm
  join together, a baby can start.

Sperm cells and egg cells and fertilization may be difficult
concepts for your child to master. In fact, it may surprise you to
know that sperm weren't identified until 1677! The ovum was-
n't discovered until even later. For most of human history, people thought that the woman was simply the incubator; the man
contributed the actual material to make babies. Your child is
likely to think so also.

Now, that may be enough of this conversation for most
preschoolers. But if your child is still interested or asks, "But
how do the egg and sperm get together?" you can introduce a
very simple definition of intercourse.

When two grown-ups love each other, they like to
kiss and hug and touch each other in ways that feel
good. Sometimes, the man and the woman place

the man's penis into the woman's vagina. The man's penis releases sperm into the woman and sometimes a baby begins.

And that's likely to be enough for even the most curious preschooler. During the elementary-school years, you can be a little more explicit. (I discuss answering questions about reproduction in greater depth in the next chapter. See pages 95 to 102.) The important thing at this age is to show your children that you are willing to answer their questions about birth and reproduction.

And what about the child who never asks? Do not assume that they have no questions about pregnancy and birth. Rather, look for those teachable moments. Pregnant neighbors and relatives, even pregnant women in stores, offer an excellent opportunity to introduce some of these concepts: "Remember when we saw Aunt Jodi last? Well, the reason her belly looks bigger is because a baby is growing inside her uterus. In a few months, you're going to meet your new cousin."

Messages for Preschoolers about Reproduction
- Both a man and a woman are needed to start a baby.
- Babies grow inside a woman in a special place called a uterus.
- Only women can have babies, but men are needed to start a baby, too.
- Mothers can feed their babies milk from their breasts or from a bottle.
- Girls can grow up to be mothers; boys can grow up to be fathers.
- Both mothers and fathers are important to children.

There are also some excellent books you can read to small children on these topics listed in the appendix.

And if you or your partner is pregnant, you have the most golden opportunity of all. Many parents have their second child while their first is a preschooler. Be sure to involve your preschool child in planning for the birth and bringing home the new baby. Bring the child with you to a doctor's visit. Share pictures from the sonogram. Talk about how excited you were when you were pregnant the first time. Talk to your preschooler about your need for nutritious foods and exercise and ask them to help you stay healthy. See the appendix for some excellent books that you can share with your preschool child during your pregnancy.

When Your Child Walks in When You Are Making Love

It seems to happen to all coupled parents sooner or later. You are making love with your partner. You have your clothes off; you are fondling, caressing, maybe having oral sex or intercourse. And you hear your door open, look over, and there is your four-year-old!

It's a good idea to be prepared for what you want to do if this happens to you. In the heat of the moment, you're likely to scream "Get out!" or some other inappropriate comment. One of my friends worriedly told me that she once rolled over after an orgasm to look directly into her three-year-old's eyes; she had no idea how long he had been standing there.

So, let's think about this situation together—in advance. First, try to imagine what your preschool child is thinking. Trust me, they are not likely to understand what is going on. This is not the time for the Big Talk on the joys of intercourse! Your child may be feeling confused, scared (is daddy hurting mommy?), or embarrassed (why don't they have any clothes on?). You may be feeling confused, embarrassed, angry, or even disappointed that you have been interrupted.

My best advice is to react calmly, take a deep breath, and gently ask your child to leave: "Sweetie, can you give us a second and go back to your room?" Then, kiss your partner, promise each other that you'll resume your intimacy at a later time, put on a robe, and go to your child. (By the way, I do not think it is a good idea to resume your sexual encounter and ask your child to wait for you. You want to deal with this one right away.)

Now, you may be asking yourself, why do I need to deal with this at all? Couldn't we just ignore that this happened? Well, for starters, you really don't know what your child saw, or how long she had been standing there, or how frightened he is. Your child may worry that you are angry or hurt. They may imagine all sorts of terrible things going on in your bedroom. And you certainly have been presented with another teachable moment.

Some parents have told me that they worry that this kind of experience will hurt their child. They worry that their child was sexually aroused or permanently confused. Freud actually labeled this situation as the "primal scene" and felt that it would harm children for life. But today, professionals do not believe that this type of single incident is harmful or damaging. (Indeed, in many societies, families all share one living space, and children observe parental lovemaking throughout their childhood without harm to their adult development. By one estimate, 75 percent of the world's children sleep in the same room with their parents.)

So what do you do? First, think about which messages you want to share. And remember, begin by first finding out what your child already knows, and if possible, what they think was going on. And try (hard) to think of this as a teachable moment: It could be an opportunity to share your values about love, intimacy, or privacy.

Here's one way the conversation could go:

Parent: Honey, you surprised us when you came into our bedroom just then without knocking. What did you see?

Child: You and Daddy were fighting. He was hurting you.

Parent: Actually, we were touching each other to show that we love each other. (*Message: One way adults show each other love is by physically touching.*)

Child: But why didn't you have any clothes on?

Parent: Sometimes, it feels good when Mommy and Daddy lie naked together. But this is only for grown-ups who love each other. (*Message: This is adult behavior.*)

Child: Are you mad at me?

Parent: No, honey. But I want you to know that sometimes grown-ups want to be alone together. When our bedroom door is shut, that means we want privacy. Next time, please knock and wait for me to answer before you come in. (*Message: In our home, we respect each others' privacy.*)

This type of conversation will probably satisfy most three- and four-year-olds. But, remember, always give your children a chance to ask more questions. And don't forget to give your child a hug at the end of this conversation and let him know that you love him.

Of course, there is a way to avoid this situation completely. Buy a lock for your bedroom door. Use it anytime you think you may make love. And teach your child that a locked door means you want privacy and that they should knock if they need you. You may still be interrupted, but at least you'll have a chance to pull yourself together.

Special Issues

Your Preschool

Increasing numbers of parents place their children in a preschool or a nursery school by the age of three or four. There are many good resources on choosing a preschool that go beyond the scope of this book, but I want to address here what you might want to know about how the nursery school addresses sexuality issues. I am not talking about whether the school has a formal sexuality education program for preschoolers, but rather how the school handles such sexuality issues as teachers touching students, toileting, and sexual play.

Preschool teachers often ask me questions such as what to do about the child who touches his genitals at nap time, how to handle the sex play going on in the playhouse, or how to respond to parents' concerns about gender-appropriate play.

You may want to ask your preschool director if the school has written policies on such issues as the following:

- Touching children. Can teachers hug and kiss children? (For a few weeks when he was three, Gregory went to a program where it turned out the teachers were prohibited from hugging children as the school's protection against allegations of sexual abuse. One day when I picked him up early, I observed a crying child, begging to be picked up, who was simply told to stop crying. I decided to look for a new preschool on the spot.)

- How are children of both genders encouraged to play together? Are there any activities that are exclusively for girls or for boys?
- How will the teacher handle questions about reproduction from a child?
- How does the school teach children about family responsibilities? Babies? Diversity?
- Does the school respect diversity in families? Are some parent/child programs in the evening so that working parents can attend? If your family is different than the mother/father/child combination, will the school support it? (For example, will your child be encouraged to make two presents for Mother's Day, or get an assignment to make a picture for Grandfather instead of Dad, etc.?)
- How are teachers expected to handle it when children touch their own genitals? In class? At nap time?
- How are teachers expected to handle sex play between children if they discover it?
- What are the policies for handling sexual contact by other children or, heaven forbid, staff?
- How does the school support parents as the primary sexuality educators of their children?

Don't be surprised or alarmed to find out that your preschool has no written policies on most of these issues. Decide which issues are important to you and what areas you want to see addressed. Ask the director of the preschool if the parents can have a meeting to discuss these issues with the staff. You might want to get the director a copy of "Right from the Start," a guide for preschools on handling sexuality issues, which can be obtained from SIECUS. (See the appendix for contact information on page 230.)

Sexual Abuse

It happens in all types of homes. It happens hundreds of times a day in the United States. An unbelievable half a million children are reported to be sexually abused in the United States each year. Seven percent of girls under twelve say that they have been sexually abused; 4 percent of boys this age say that they have been sexually abused. In 90 percent of the cases, the assailant was someone in the child's family or someone close to the family. Girls are more likely to be sexually abused than boys, and men are more likely to be the assailants than women. But boys are also victims of sexual abuse, and some abusers are women. And sexual abuse is every parent's nightmare.

The widespread child abuse scandal in the Catholic Church during the past few years should be a wake-up call to parents. We have been horrified by the number of priests who have now been convicted of child sexual abuse. The horrifying truth is that there have been reported cases of child molestation by religious leaders of many denominations. People who molest children don't look like deranged monsters; they work with us, go to church with us, minister to us, and live with us.

The statistics are alarming and overwhelming. In another study, 16 percent of teen boys and 27 percent of teen girls reported that they were sexually abused before the age of twelve. Eighty-five percent knew the person who abused them. And despite the highly publicized cases of sexual abuse in day-care centers, only 2 percent of reported cases of abuse happen there. Frighteningly, most children are sexually abused in their own homes.

I can imagine that many readers will want to skip this section. But it is important for every parent to know the signs that might indicate child sexual abuse and know what can be done to prevent it.

Most parents know to teach their children not to talk to strangers, accept candy from people they don't know, or ever get in a car with a stranger. Indeed, many preschools and kindergartens offer "stranger danger programs." And although this is obviously good advice, the reality is that most people who abuse children are known to them.

One woman I know came home and found her husband fondling her two-year-old's genitals. A friend of mine came home from work early and discovered the baby-sitter inserting her finger in and out of her daughter's vagina. Many of my adult friends have stories of an uncle who fondled them when they sat in their lap; a friend's father who exposed himself to them at a slumber party; a youth worker who invited them to their home and then tried to get them to look at pornographic magazines.

Children can also be sexually abused by other children. This is different than the sex play I described earlier in this chapter. In sex play, children are usually of the same age, are engaging in lighthearted exploration, and appear to be enjoying themselves, at least until they realize they have been "discovered" by an adult. And, in general, once they are told by an adult that they should stop, they do. (Turn back to the chart on page 60.)

Some children, however, engage in more inappropriate behaviors and become sexually abusive to other, usually younger, children. This may include engaging in oral or genital sex with other children, with or without their consent. Many of these children have been sexually abused themselves, and many of them have inappropriately been exposed to sexually explicit materials or adult erotic behavior. In one study, all of the girls who had sexually abused other children (about 25 percent of childhood abusers are girls) and most of the boys had been molested themselves, often years before they committed these crimes against other children. In a case reported widely in the news, three boys in Dallas aged eleven, eight, and seven are being charged with abducting, beating with a brick, and sexually assaulting a three-year-old girl whose mother went into the house to make a phone call.

So, how do you protect your child from sexual abuse? First, I think it is important to understand that there is no 100 percent foolproof way to protect your children. My friend's world was ripped apart when she discovered that the man she had married was a pedophile (a pedophile is a person who is sexually attracted to and seeks sexual behaviors with children). References for baby-sitters may not know about a person's sexual desires. Neighbors and relatives can seem perfectly trustworthy. As our children grow up, we often entrust them to teachers, ministers, priests, camp counselors, and scout leaders. And although it is frightening to contemplate, many pedophiles choose to go into these youth-serving professions primarily to have easy access to children.

I do not believe that children can protect themselves from sexual abuse. Many of the programs that are conducted in preschools and kindergartens say that they are sexual abuse *prevention* programs. The reality is that they can't prevent child sexual abuse from happening. They *can* help protect children by helping them to identify child sexual abuse and know what to do if they are being abused.

"Good touch, bad touch" programs try to teach children the difference between appropriate and inappropriate touching; my experience is that these concepts are very difficult for children who are younger than the ages of seven or eight to understand. Sexual touching by an adult may feel like "good touch" to the child, and being examined by the doctor or having their hair washed may indeed be perceived as "bad touch." These labels simply do not work for most preschool-age children.

"No, go, tell" programs are better: They teach children to say no to the abuse, leave the situation immediately, and tell a parent or caregiver if someone tries to hurt them. But both of these programs are based on the assumption that the child has the social or physical power to stop an adult pedophile's actions, something that is unlikely to be true. Further, many of these programs are surprisingly squeamish about using correct terminology for the body; most of them teach children about their "private parts" or "the parts of their body covered by a bathing suit," once again teaching the idea that adults feel the genitals are somehow negative or bad.

These programs may also give children the wrong idea about what is appropriate touch. My mother still remembers the day that she was giving three-year-old Alyssa a bath, and Alyssa started to scream, "Don't touch my vulva!" when my mother passed a washcloth between her legs. Alyssa had been taught in preschool that week that *no one* should ever touch her genitals. And a friend of mine called me after her five-year-old complained about her patting him on the buttocks, "Mom, I'm going to have to tell my teacher you did that."

I do not mean to imply that these programs don't have a place, just that you should not expect them to safeguard your children. They can give your preschooler some important messages about sexual abuse. And these programs do seem to help children tell someone if abuse has occurred. But it is your job as a parent to keep your child safe.

I believe that parents can prevent *some* childhood sexual abuse. I was once asked by a journalist who was writing an article for a women's magazine what was the one best thing a woman could do to prevent her children from being sexually abused. I answered, "Marry the right man." For the reality is that about three-quarters of both boy and girl children who are sexually abused are abused by their own fathers, their stepfathers, or their mother's boyfriends. And since most people who sexually abuse children were sexually abused themselves, it is important that adults who have been abused thoroughly deal with this sad part of their history in therapy.

You must also be very careful about screening nannies, preschools, and child-care providers. When your child gets older, you should ask whether such community-based organizations as Scouts, Little League, and the religious education program at your church, synagogue, or mosque, perform background checks on potential employees. I have asked every potential nanny applicant that I have hired on the written application, "Have you ever been sexually abused?" I have always done police checks on my nannies to make sure there were no cases pending against any preschool or child-care center we were considering (you can call your state's Department of Social Services to check this out; you can also just ask the school.) References are critical for baby-sitters, nannies, and child-care workers. You can also not allow your children to play at other children's houses until you feel you know the other parents well.

As a minister, I am especially concerned that our churches, synagogues, and mosques are safe places for children. If your child is attending a religious education program or youth group, you can check to see that your religious institution takes these issues seriously. Do they have a screening policy, which includes background checks for all staff and volunteers who work with children, youth, and vulnerable adults? Do they have annual training on child sexual abuse prevention? Are there requirements that every classroom have two adults present? Is there a written policy on how to respond to a convicted sex offender who wants to come to church?

Despite all of these precautions, child sexual abuse can still occur. Here are some behaviors that should raise concerns about the possibility of sexual abuse:

- Your child has an unusual discharge from their penis or vagina. Call your doctor for an appointment today. It is most likely just an allergic reaction to new soap or bubble bath, or keeping them in a wet bathing suit too long, but it could also be a sign of a sexually transmitted disease. Your pediatrician can tell the difference.
- Your child is compulsively masturbating in public, after you have repeatedly told him or her that this is private behavior.
- Your child tries to get other children or adults to touch his or her genitals or repeatedly tries to touch your genitals after being told to stop.
- Your child begins to be more interested in sex or engaging in sexual behaviors than playing with friends, going to school, or other activities.
- You discover your child engaging in sex play with children who are several years older than they are.
- Your child manually stimulates or has oral or genital contact with your pets.
- Your child repeatedly draws pictures with the genitals as the primary focus.
- Your child begins to exhibit disturbing new toilet behaviors, such as playing with their feces.
- You discover your child engaged in oral-genital sex with another child.

This is a rough list. Any parent would be horrified to discover their child engaging in any of these behaviors or problems. But, if you do, try to stay calm. Call your pediatrician and ask for an immediate appointment. Ask for a referral to a mental-health professional who specializes in child sexual abuse and assessment. Seek counseling and support for your child and for you. If your child has been abused, seek specialized treatment programs; they can help children who are sexually abused and children who sexually abuse others. Call the National Clearinghouse on Child Abuse and Neglect Information, (800) FYI-3366, for help, referrals, and advice.

If you were sexually abused as a child, I know that this section may have been especially difficult for you to read. If you have not dealt with this part of your past, you need to know that you were not responsible for the abuse and that many childhood sexual abuse survivors feel shameful, guilty, or depressed. There is effective treatment and hope. Call the Safer Society Foundation Inc., (802) 247-3132, or see their Web site, www.safersociety.org, for information and referrals.

It is also important that you not let your concerns about sexual abuse affect how you allow your children to be touched. You want your child to be held, kissed, and cuddled by the caring adults in their lives. You want them to experience touch as a good thing. And you want to teach them that they can control who touches them.

Children seem to know by instinct whose touch is comfortable and safe. They hate the aunts and uncles and grandparents who squeeze their cheeks too hard or who smell funny or don't respect their boundaries. I believe it is important to allow children to accept and refuse touch, even from relatives. Yes, it's hard on you if little Christine doesn't want to kiss Grandpa, but it is never a good idea to force your child to hug or kiss or cuddle anyone. This drives my own mother crazy; she believes that our children should say good-bye to all of our relatives with a hug and a kiss. But it is more important to me that our children know that we respect their feelings and that they have the right to say no to any unpleasant touching than it is to disappoint one of our relatives.

Staying involved in your children's lives as they grow older is one other way to protect them from sexual abuse. Don't just drop them off at Scout meetings or soccer practice or church school; hang around, talk to the leaders, get to know them, and ask the organizers how they were selected. Watch out for adults who seem *too* interested in your child; requests for time alone with your child from people you don't know well may be indications of potential trouble.

Messages for Preschoolers about Sexual Abuse

- Your body belongs to you.
- There are good reasons for some adults to look at or touch children's bodies, such as a parent giving a child a bath or a doctor or nurse examining a child. It is okay for a parent (nanny, other child-care provider) to help you wipe yourself after you use the bathroom.
- Come to me if any other person makes you feel bad or funny or does something that makes you think, "uh-oh."
- You can tell someone not to touch your body if you don't want to be touched.
- If someone touches you and tells you to keep it a secret, tell me anyway.
- If a situation makes you feel funny or bad, say "no," try to get away, and tell me right away.

Values Exercise for Chapter 4

You are watching a TV program with your seven-year-old, and a female character makes a sexually suggestive comment to a male character. You...

- ❏ a) do nothing and hope your child missed it.
- ❏ b) turn off the TV and say, "You can't watch this anymore."
- ❏ c) say, "What do you think she meant by that?"
- ❏ d) wait for the commercial and say, "This show makes me uncomfortable because I don't like the way the adults treat each other. What do you think?"

Your six-year-old son wants ballet lessons (or your five-year-old daughter wants to play touch football). You...

- ❏ a) say, "Only girls/boys play football/take dance lessons."
- ❏ b) say, "Okay, let's look into them for you."
- ❏ c) say, "Tell me more about what interests you"—and then suggest a more gender-specific substitute.
- ❏ d) change the subject.

Your child gets into the car after school, and says, "Oh ****, I forgot my backpack." You say...

- ❏ a) "where did you learn that word?"
- ❏ b) "don't you ever talk like that to me again."
- ❏ c) "you sound really upset. Let's go get your backpack together."
- ❏ d) "let's get your backpack and when we get home, we'll talk about the word you just used."

You discover your six-year-old and two neighborhood children playing "Tag, you have AIDS." You...

- ❏ a) ignore it and let them finish playing.
- ❏ b) stop them and sit them down for a talk about AIDS.
- ❏ c) talk to your child later about AIDS and ask her not to play tag this way again.
- ❏ d) ask, "Where did you hear about AIDS?"

Chapter 4
Early Elementary School
Ages 5 to 8

I cried the first day that I took Alyssa to kindergarten at our local public school. I walked her into her classroom, introduced her to the teacher, kissed her good-bye, and burst into tears as I walked to the parking lot. She had been in a private child-care center since she was three; it was not separation anxiety that caused my tears that morning. It was my sense that she was leaving the world where we could tightly protect her and entering the wider world where we could not. She would be exposed to children and grown-ups with different values, to teachers who might not be as demonstrative and caring as the ones in her preschool, to a much larger number of students, and to situations in which her feelings might be hurt and her beliefs challenged. I had a strong instinct to ask the teacher for a private meeting so I could tell her how very special our child was and why she needed her extra-special attention. Home schooling looked pretty good to me at that moment!

Your Child Is Growing Up

The early-elementary years are years of tremendous growth and change. Naps and playtime give way to homework and tests. Preschool hugs may be replaced by colder hellos. The class of twelve now has twenty children, and individual attention is lessened. Your child discovers that not everyone lives the way you do in your home, and older children are a much stronger influence. Your child will hear more than you would like about sexuality issues on the bus or the playground from other children.

Freud wrote that the period from six to puberty was a latency period for sexual issues. He wrote that children this age simply aren't interested in sex: They are too busy growing socially and intellectually. And this out-of-date theory continues to be used to fight sexuality education for elementary school–age children.

Nothing could be further from reality than this concept of latency. Your five- to eight-year-old continues to be very interested in sexual issues; it is just that it is no longer as apparent to the adults in their lives. Some professionals believe that latency is mostly a result of a sexually repressive culture; in cultures where sexuality is openly expressed, children continue to express these interests throughout childhood. In a recent study of more than 11,000 children ages two to twelve, mothers reported that they had observed sexual behaviors in children at all of these ages. For example, 14 percent of six- to nine-year-old boys were still touching their genitals in public, 40 percent did so at home, 20 percent tried to look at people nude, 8 percent wanted to watch nudity on TV, and 14 percent were very interested in girls. Girls this age exhibited sexual behaviors as well that were observed by their mothers: Twenty percent had been observed touching their genitals, 20 percent tried to look at people nude, 8 percent wanted to watch nudity on TV, and

14 percent were very interested in boys. And that's just the behaviors that the mothers were able to observe!

Your five- to eight-year-old is continuing to develop as a sexual person. Five- to eight-year-olds are very curious about pregnancy and birth. They are developing stronger friendships, and most boys and girls show a strong preference for playing with children of the same gender. They are becoming even more aware of societally defined gender roles; they have a clearer idea about what is expected of boys and what is expected of girls. They may continue to engage in sex play with children of both genders, although they are much more likely to do it where they will not be discovered by adults. And in private, their exploration of their genitals may become more purposeful.

Some sexologists even believe that the period from ages five to eight are critical years in sexual development. According to the late Dr. John Money of Johns Hopkins University, each person develops a love map during the first years of life, similar to the way we develop a native language, and, he says, love maps are completed by around the age of eight. Dr. Money defined love maps as "a developmental representation synchronously existent in the mind and the brain, depicting the idealized lover, the idealized love and sexual affair, and the idealized solo or partnered program of sexuerotic activity projected in private imagery and ideation or in observable performance."

In other words, our love maps develop in our brains, and help us develop a picture of our idealized romantic and sexual partner, including their preferred build, race, color, temperament, and look, as well as developing our preferred sexual behaviors in adulthood. According to Dr. Money, by the age of eight, these love maps are pretty well established.

And many professionals believe that the elementary-school years are the most important time for people to develop as moral thinkers, an important part of adult sexual health. While they are preschoolers, children believe that their way of thinking is the only way possible. But in early elementary school, children begin to understand that there may be other points of view and ways to consider a situation. They can begin to understand the "golden rule": Do unto others as you would have them do unto you.

Developing an ability to empathize and make good decisions is part of the foundation for adult sexual health. A sexually healthy adult is able, for example, to make decisions about their sexuality that are consistent with their own values, and can discriminate between behaviors that are life enhancing and those that could be self-harming or dangerous to others. Giving your elementary-school child plenty of opportunities for decision making ("Would you like to wear the blue shirt or the green shirt today?" "What would you like in your lunch?") gives them the chance to practice this skill. Exploring with them possible alternative endings to stories helps them think about different possibilities: "Instead of marrying the prince, what else could Cinderella have done?" Reading them books and stories with moral messages provides an opportunity to talk about important subjects: "How was Wilbur being a good friend to Charlotte?" And helping them understand that all decisions have consequences teaches them that their actions affect themselves and others.

Five is a good age to begin the practice of "family meetings." Some families get together once a week to discuss issues of family concerns. Others do it when there is an important family issue to be discussed or decided. Meetings can be a good opportunity for children to help make family decisions. For example, children can be involved in deciding such issues as what activities to do on the weekend, where to go on a family vacation, and if the family should get a new pet. Look for opportunities to involve your child in family decisions. Family meetings can also be used for individual family members to bring up issues and concerns, including everything from household chores to new baby-sitters.

Be prepared, though, that your little decision maker may want to be involved in some things that aren't appropriate. When we told six-year-old Alyssa that we were going to have a new baby, she angrily asked, "How come I wasn't included in the decision making?" Sometimes, we told her, adults need to make private decisions without their children's help! It did, however, provide us with the perfect "teachable moment" to talk with her about how babies are made.

Messages for Early Elementary–Age Children about Decision Making
- Everybody has to make decisions.
- All decisions have consequences.
- Decision making is a skill.
- Children need help from adults to make some decisions.
- Parents can help children make decisions.

Teaching Your Child the Facts of Life

It happens to many elementary-school children. An older child, perhaps a fourth or fifth grader, asks them on the bus or on the playground, "I bet you don't know where babies come from" or far less nicely, "Do you know your parents f***?" This child then proceeds to give the younger child their interpretation of reproduction and intercourse. And worse of all, most parents will probably never know that this conversation even took place.

Stop and think for a moment how you learned about intercourse. Who told you? How old were you? What was your initial reaction? Could you believe that your parents had done this? Did you believe the person who told you? When do you remember having the first discussions with your parents about sex and sexual intercourse? (Some people have told me that it was the night before their wedding!) Many of us were told about intercourse for the first time by an older friend or sibling, and most of us were pretty shocked, maybe even repulsed. We thought, "My parents didn't do that" or "My parents only did that twice, once for me, and once for my brother." Some of us still like to think that!

I believe that ages five to eight are the right time to introduce the concept of intercourse. I am not talking about giving your children detailed information about lovemaking. But, I am talking about making sure that your child can answer the older child on the playground with "My mom and dad talked to me about this; I'm not going to listen to you."

You may be asking yourself, Do I really need to have this talk with my child? Wasn't learning it on the street good enough for me? Well, was it? Think about some of the myths that some of us first learned: The stork brought you, you grew in a cabbage patch, you get babies at a store. One of my colleagues told me that his mother still remembers his response to being told that "the father plants a seed in the mother." My colleague, at five, retorted, "With a shovel?"

Try to remember the first time someone told you about intercourse: Did you really believe that these two outstanding people you called your parents could do something like that? Think about an older child's giving your child an ugly impression of adult lovemaking rather than your giving them a caring, sex-positive message that emphasizes that this is an adult behavior. And if you still don't agree that this is the time, think about having your first conversation about intercourse when your child is a smart-alecky preadolescent who says, "I already know all about this." (Of course, if you really don't agree, please skip ahead to the next section of this chapter.)

Psychologist Anne Bernstein says that early-elementary-school children move from "geographers" to "manufacturers" in their questions about reproduction: They are less interested in *where*, and more interested in *how* babies are constructed. They are unlikely to be satisfied with answers only about sperm cells and egg cells, and they are likely to now want to know how the sperm and the egg get together.

Many adults have told me that, in the absence of a parent offering them a definition of intercourse, they figured out their own ways to join the sperm and the egg. Maybe it happened when a woman sat on a man's lap. Maybe sperm swim across the bed when grown-ups are sleeping. Maybe it happens in a swimming pool. And then they recount that, as small children, they were afraid to sleep with their brother, or sit on their dad's lap, or go to the public pool because they didn't want to get pregnant. One of my colleagues shared with me that in answering this question, he told his five-year-old son, "The mother and father lay very close together." After a moment of being puzzled, his son replied, "Dad, *that* doesn't explain anything!" After my colleague stopped laughing, he agreed and gave him a little more detail.

So, what do you say and when do you say it? Let's think about some possible teachable moments: You see a pregnant woman in the park, you read a book together about pregnancy and birth, there are two adults moving toward a bed on a TV show you watch together. Here is one way the conversation could go:

Parent: Do you remember when we talked about how babies are made?

Child: Yeah.

Parent: What do you remember?

Child: Something about the sperm and the egg.

Parent: That's right. Well, do you also remember that you need a man and a woman to start a baby? In the man, there are special cells called sperm; in the woman, there are special cells called ova. When a sperm cell and an egg cell get together, sometimes a baby can begin. How do you think the sperm and the egg get together?

Child: Does the sperm swim across the bed?

Parent: That's a good guess. But what really happens is that when grown-ups love each other, sometimes it feels good when the man and woman place the man's penis inside the woman's vagina. After a while, the man's sperm come out of his body and travel up the woman's vagina to her uterus. Sometimes the sperm and the egg come together inside the woman and that's the beginning of the fetus that will grow into a baby.

And then wait. See if your child can absorb this information. See how they react. Wait and see if they have another question. You may want to share with them some of the books on babies for children ages three to eight listed in the appendix.

One possible reaction your child might have is "Oh Dad, that's gross," and look disgusted. This is not an unexpected reaction when a child first learns about intercourse. It may seem impossible that the penis could fit into the vagina or the thought may be just too bizarre. This is why it is important that a parent or caring adult introduce the idea of intercourse, not a playground buddy. You could say, "I understand that this seems yucky to you. That's okay, because this is something that only grown-ups should do. When you're a grown-up, you'll make up your mind whether you like it or not. We can talk more about this at another time." And then give them a big hug (and if you must, go hyperventilate in another room!).

Your children could also have some other reactions. One of my friends reports that his son asked if he could watch the next time. Another friend told me that her daughter asked her if she (her mom) could be with her (the daughter) the first time she had intercourse. Still another told me that her daughter said, "No, Dad, I heard that you buy babies at the store." Someone recently told me that her four-year-old said, "Mom, that's marvelous!" No matter what your child's initial reaction is, you will want to go through the same steps: respond to your child's feelings, give correct information, give your values, and keep the door open for future conversations.

There are some other points that I think are important in introducing the concept of intercourse to your child.

It is important to get across to your child that sexual intercourse is also a way to demonstrate love and pleasure, and it is not just for making babies. After all, you don't want your child to think that people only have intercourse to get pregnant (unless that is your family value), and I do not think it's sexually healthy to grow up thinking that your parents only had sexual intercourse the two or three times when they conceived you and your siblings. (You might be surprised by how many of the adults I know still don't think—or at least do not want to believe—that their parents had sexual intercourse except to reproduce.)

By the way, I don't recommend using animals as the way to introduce intercourse. I never have understood why people refer to sexuality education as the "birds and the bees"; somehow pollen and nectar never did help me understand human reproduction or sexual pleasure. Taking your child to the zoo to see monkeys copulate probably is not a good idea either. Watching animals copulate can be confusing and scary to children unless, like on a farm, it's a regular part of their environment. (For example, a friend of mine breeds llamas, and her daughter has been around the llamas breeding since she was quite small. When she reached early elementary school, they were able to use inseminating the llamas as a teachable moment. However, this wouldn't work for most of us.)

You can also introduce an early-elementary-school child to the concept that people can choose whether or when to have a child and then, how many children they want in their family. Five- to eight-year-olds can learn some simple information about contraception. They can understand that all children should be wanted, that some families have children and others do not, and that each family can decide how many children to have. You can introduce the idea that there are ways to stop the sperm from meeting the egg: it is called birth control.

Introducing the concept of contraception at this age teaches your child that sexual behaviors have consequences, and that intercourse must be protected. You don't need to make a big deal of this; you are likely to have situations where this will arise naturally. For example, if you are going to the store with your child and need to buy contraception or condoms, don't put off buying them for an additional trip when you are alone. If your child asks, tell them that you and your partner love each other and the children, but have decided not to have any more babies right now. One of my favorite research studies shows that teenage girls who know which birth control method their mother uses are more likely to use a birth control method themselves. Introducing the idea of contraception at the same time you introduce the idea of intercourse gives young people the important message that caring sexual experiences are protected.

Years ago, I was on a study tour to Sweden and had the opportunity to meet with several groups of Swedish teenagers. Swedish teenagers begin sexual intercourse at about the same time that Americans teenagers do, but they have one-sixth the pregnancy rate. When I asked a group of Swedish ninth graders if they would ever have intercourse without contraception, they looked at me as blankly as if I had asked them why people might kill each other. I explained to them that in the United States, many teenagers don't use contraception. They were still silent. One of the preteens finally volunteered, "But that would be silly unless you wanted to become pregnant. It would be like driving through a red light." And they explained that they had known about using birth control as long as they had known about having sex. I believe that this is the attitude we should be striving for in the United States.

I also believe that it is important not to overemphasize the relationship between conception and marriage. I personally think it's better to talk about "grown-ups" than "married people" because your child is soon to meet, either in person or through the media, adults who have children who aren't married, if they haven't already. This is not to say that you shouldn't teach your child that you believe that only married people should have babies if that is your family's value. Just separate your values from the fact presentation. For example, you could say, "Some grown-ups have children without being married. In our family, we believe..." (and then state your value about out-of-wedlock childbearing).

These simple exchanges can begin to give your child a healthy introduction to sexual intercourse. Answering their questions or providing them with simple information can convey to your children that (a) you respect their feelings; (b) intercourse is for grown-ups; (c) adults make decisions about when and whether to be involved in sexual behaviors; and (d) in your home, you talk about sexual issues. Not bad for a three-minute discussion!

Messages for Early Elementary–Age Children about Anatomy and Reproduction

- Male and female bodies are equally special.
- Reproduction requires both a man and a woman.
- Men have sperm cells and women have egg cells in their bodies that enable them to reproduce.
- Intercourse is when a man and a woman place the penis inside the vagina.
- Intercourse is for love, pleasure, and making babies.

Bath Time

A few years ago, I received a worried call from a colleague. He was a doctor specializing in adolescent medicine, and he was the parent of a six-year-old son and a five-year-old daughter. We were talking about a work issue, when his voice got quieter and he asked me if he could ask me something personal. "My wife and I have decided our children should stop bathing together, but they don't want to. We're worried that they may start having sex together."

I calmly asked, "Why?"

He answered, still in a worried whisper, "Because they'll see each other's bodies and want to experiment."

I asked him, "I hear that you are worried about this, but I'm wondering if they have ever done that? Do you have any reason to suspect that they want to have sexual contact with each other?"

He sheepishly admitted that he had no real reason to think about this, but that it was a worry and a concern. I suggested to him that he take off his adult lenses: Yes, his children were likely to be curious about each other's different body parts, but it was highly unlikely that they would be erotically aroused by the sight of each other.

And indeed, joint bathing provides another teachable moment. I suggested to him that if his children were comfortable bathing together in early elementary school and seemed to want to continue, that he and his wife reconsider. They should be sure to tell their children not to touch each other's genitals because those parts of the body are private. They could also let them know that if either of them become uncomfortable bathing together, they should tell a parent and it would come to an end. And to make themselves more comfortable, the parents could leave the door open and pop in now and then, a good idea also to make sure the children are safe and indeed are getting clean!

I also told him that the six-year-old was probably going to be interested in giving up this activity pretty soon. Shared baths were likely to give way to private showers. Even in homes where nudity is common, children seem to go through a "modesty stage" in which they are no longer willing to change in front of their parents or walk around without clothes on. I can still remember how embarrassed I felt at age five when my mother wanted to change me from a wet bathing suit to a dry one on the beach; I wanted the privacy long before she recognized that I needed it.

It is important to respect your growing child's desire for

privacy. It helps if they have a bedroom door they can close and that you agree that any member of the family will knock before they enter a room with a closed door. I've given my children the little plastic signs from hotels that say, "privacy please," and told them to hang them outside their door when they want to be alone. We will knock and wait for them to answer before we come in. These discussions about privacy also help set the stage for how to deal with your child's more deliberate exploration of their genitals at this age.

Masturbation

Genital self-touching may become more deliberate during the early-elementary years, or it may not. And there probably isn't much you can do about it. One of my women friends remembers having her first orgasm at around six or seven. One of my colleagues tells this story about her six-year-old daughter. As a small child, her daughter often rubbed her vulva against furniture and dolls, and they had taught her that this behavior needed to be done in private, in her own room. When her daughter was about six years old, her mother entered her room one evening to kiss her daughter good night. She realized by her daughter's noises that she was masturbating. Her six-year-old daughter said, "Mom, could you go out? I'm rubbing and I was just getting to the good part." Clearly, she had discovered the joys of self-pleasuring.

At this age, it is normal for children to masturbate, and it is normal for them not to. In one study, researchers found that 40 percent of the boys and 20 percent of the girls ages six to nine masturbate with their mother's knowledge. And that's just those children whose parent knows the behavior is going on; in fact, the incidence is likely to be much higher, as many six- to nine-year-olds have learned that this is behavior best kept hidden from adults.

Just as with your younger children, you need to think through which values you want to give your children about this behavior. If your family thinks masturbation is wrong, you should tell that to your child. If your family thinks masturbation is a healthy private way to experience sexuality, your child should know that. But all children, regardless of your family's values, do need to know that masturbation doesn't cause physical or mental harm, and that it should be done in a private place. Your early elementary school–age child may also be exploring these feelings and curiosity with other children.

More on Sex Play

Sex play continues through the early-elementary-school years. In fact, for some children, it begins at this age as it may be their first real opportunity for unsupervised play. The major difference between sex play for five- to eight-year-olds and sex play during the preschool years is that by second or third grade, most children take better care to ensure that they won't be discovered.

My husband remembers that there was a big pine tree in his neighborhood where children used to go to undress each other. In the neighborhood I grew up in, one of the kids lived in a house with an unfinished basement that parents never seemed to enter. A friend of mine remembers sex play with neighborhood children in one child's tree house.

Let me tell you about what happened to one of my friends. Her third-grade daughter came home from a play date with a male classmate. When my friend asked her daughter what they had done, Tracy said, "We played a game. Mom, if I tell you something, do you promise not to get mad?" My friend said, "Of course." Tracy then went on to say that she and her friend Rob had played a game and the loser had to do anything the winner said. She said that when she had lost, Rob told her to take off her pants and he would kiss her "privates." My friend, trying hard to stay calm, asked her, "Well, what happened?" "Well, I lost, I did it, and I kinda liked it." My friend, barely breathing at this point, then said, "What happened next?" "Well," Tracy said, "when Rob lost, I told him to take down his pants and I would kiss his penis. Are you mad, Mom? Rob told me not to tell." My friend, not knowing what to do or say, and not wanting to violate the trust her daughter had placed in her, managed to get out, "This isn't behavior that children should be involved in. Please don't do it again."

After she had some time to think about it, she decided to call Rob's parents. They didn't believe her! Unbelievably, a week later, they called to see if Tracy could sleep over at their house that weekend. Our friends said "no," and put an end to Tracy and Rob's having any unsupervised time together.

What was going on here? Well, looking at the chart on page 60 in the last chapter, this could have been simple childhood curiosity as it obviously was on Tracy's part or perhaps it was an indication of something else on Rob's. Clearly, Rob had been exposed to age-inappropriate information about oral sex; the question is whether it was through personal experience in an abusive situation or more innocently, perhaps in overhearing an adult conversation or watching an adult cable show or movie.

The question that parents need to think about is whether they feel comfortable with this type of play, whether they want to do everything possible to make sure it doesn't happen, or something in between. Many parents mistakenly assume that this is just about children of mixed genders playing together; it is not. Sex play is just as likely to be between children of the same gender as it is among children of different genders, and it is not related to a child's future sexual orientation.

Some commonsense rules will reduce, although not eliminate, the possibility of childhood sex play in the early-elementary-school years. You can establish a family rule that visitors must be entertained in family spaces like the den or living room, or that bedroom doors must be left open. You can tell your children that no one else should touch their genitals and that they shouldn't touch anyone else's body either.

And what do you do if you discover this type of play in the elementary-school years? Take another look at the chart on page 60. If the behavior falls in the expected column, stay calm, state your values, and lay out your clear expectations for future play. However, If the behavior is more troubling, you may want to consult your pediatrician or a mental-health professional. And if you think it is possibly sexual abuse, please go back to the section on pages 82 to 89 to think through how to handle this situation.

Children experiment with all types of friendships during the elementary-school years, and they need your help in learning how to be a good friend.

Friendships

During the early-elementary-school years, your child is learning more about being a friend. While during the preschool years, most friendships are based on convenience (play dates arranged by parents, other children at the preschool, next-door neighbors), during the first years of elementary school, children begin to select their own friends. Most girls, on average, tend to develop a few "best friends" at this time; most boys, on average, have a greater number of friends but less-intimate friendships.

This is also a time when some children, for a variety of reasons, begin to be rejected. It is important for you to know if your child has friends. Unpopular children often feel sad and rejected, and suffer from low self-esteem. I confess that this is a very personal issue for me: My entire second-grade year was ruined by the "Hate Debbi Haffner" club. Neither my teacher nor my parents intervened. And although by third grade, this was over, it is a memory that I have never forgotten.

If your child seems to be having difficulty making friends at school, talk with the teacher. Observe your child in class. Invite children over to play. Talk to your child about what makes a good friend. They can learn how to have a good conversation, how to show interest in other children, and how to initiate friendships.

For some children, it can help to provide alternative opportunities to make friends: Sign your child up for a sport, a Boy or Girl Scout troop, or a program at your church or synagogue. Invite neighborhood children over for play dates, or have your friends come over with their children. Try to make sure your child has some pleasant opportunities to play with children of their own age.

And if you discover that it is your child who is ostracizing other children, talk to them about how important it is to be open and caring to others. Help them develop empathy by talking to them about how it would feel to be teased or rejected. Let them know that you will not tolerate their being mean to other children. Often children who are different in some way—those with disabilities or those who don't fit expectations for their gender—are most likely to be bullied.

Tomboys and Sissy Girls

I was driving Greg and three of his boy friends on a school field trip. One said, "So, do you think the new girl is really a girl?" Another said, "How come she dresses like that and likes playing with us better?" I listened, realizing that they had seemingly forgotten that I was in the front seat. They weren't really being mean, just curious about this girl in her army pants, short hair, boy-like sneakers, and complete lack of interest in "girl things." As we pulled into the parking lot, I said something like, "Every person has different likes and dislikes. I hope you will play with this girl. It sounds like she wants to be your friend."

Unlike the preschool boy who for a few weeks wants to try on high heels and play with dolls, or the preschool girl who won't wear a dress and only wants to play with trucks, by elementary school, children are much more aware of society's expectations for their roles as boys and girls. Yet, some children

---

Messages for Early Elementary–Age Children about Friendship

- People can have many friends.
- A person can have different types of friends.
- Friends spend time together and get to know each other.
- Friendship depends on honesty.
- Friends can feel angry with each other.
- Friends can sometimes hurt each other's feelings.
- Friends forgive each other.
- Friends share feelings with each other.
- Friends can help each other.
- Friends can be either male or female.
- Friends can be younger and older.
- It is wrong to be mean or cruel to another child.

---

persist, despite social pressure to conform, to identify more strongly with children and activities of the other gender. According to the National Children's Medical Center, "gender variance and gender non-conformity refer to interests and behaviors that are outside of typical culture norms for each of the genders."

Girls who prefer to act like boys are often labeled "tomboys," and parents are often less alarmed by a daughter who wants to play football than a son who wants to study ballet. A tomboy may be more accepted by her classmates, than a boy who is seen as a "sissy." But, both are likely to face teasing and social isolation, especially if their outward appearance seems more like the other gender than their own.

It can be difficult for parents to have a child who expresses gender-variant behaviors, especially parents with strict gender-role expectations. The fact is that these children do not choose to have gender-variant interests any more than other children deliberately choose more socially expected gender behaviors. Many experts today believe that these differences among children are mainly the result of biological influences and cannot be changed by parents, teachers, or therapists, any more than left-handedness can be changed by insisting that the child write with the right hand. Yes, outward behavior change can be forced, but inside, the child will probably hide their real interests and feel rejected by those who are insisting they be different than they are.

If you are troubled by your child's gender-variant behavior, it may help you to talk with other parents who have had this experience or a therapist with training in these issues. This may be especially important if you and the other parent disagree on how to address these issues with your child. A therapist can help you and your child learn to cope with intolerance, bullying, and prejudice, and a support group of parents with similar children can help you not feel alone. Be wary of therapists who want to force your child to adopt behaviors more typical of their gender. See the appendix for "Outreach Program for Children with Gender-Variant Behaviors and Their Families" for referrals and an electronic support group.

The most important message we can give our children is that we love them, just the way they are. I was once approached after a talk by an elderly couple. The grandfather took out a picture of a seven- or eight-year-old boy. "This is our problem," he said. "His brother is all boy. Well, he's not. He likes to play with dolls, he is very gentle, and he only seems to like girl things." I told them that children do not choose to be different, and that this little boy needed their acceptance, support, and help dealing with what were probably difficult reactions at school. I told them to encourage him to become involved in non-gender-specific activities like swimming and karate, where both boys and girls participate. Grandpa was a bit agitated; Grandma listened quietly. I told them that at some point this little boy might need counseling to help him deal with discrimination or intolerance, but right now what he needed most was for them to love him unconditionally. Grandma tapped Grandpa triumphantly on the arm, "See, that's just what I've been telling you."

Going to Other People's Houses

When your child was in preschool, you were probably arranging all of their play dates and probably going with them to play at other people's houses. Now that they are in elementary school, they are starting to be invited to homes where you may not even know the parents. And they may even be invited to sleep over at these homes. In essence, you are being asked to give your child permission to be taken care of by an adult you hardly know.

I found this quite challenging when it first happened. It is our job as parents to make sure Gregory and Alyssa are safe; was I really going to allow Alyssa to go someplace I didn't know? What kind of values might she be exposed to that were different than our family values? What kind of television or videos could she be allowed to watch? Did this family place the same priority on child safety that we did—were the medicines and cleaning supplies locked up high away from little children's reach, or were there firearms in the house? And was there any possibility of her being sexually abused? (Remember, children are most likely to be abused by people they know.)

I suggest being conservative in this area until children reach adolescence. You could decide that your child may not go into another child's home, unless you know the parents and have given them permission to play there. That means that you have to get to know the parents of the children your child is becoming friends with. Invite them all over for a family play date; your children can play while you are getting to know each other. Go with your child to their first play date at another child's home. Spend time getting to know the parents. Talk openly about your family rules on television and videos. Come up with agreements on how your children can play together. ("I'd prefer that they only play with adult supervision" is one example.)

And make sure that you reinforce some new, more grown-up messages with your child about sexual abuse prevention. Let your child know that they should tell you if anything uncomfortable happens when they are on a play date. Be sure to spend a little time at the family's home before or after every play date. Ask your child to tell you what they did. Cultivate the parents of your children's friends as your friends. Agree on rules about such things as time alone, unsupervised outside play, and television watching.

> **Messages for Early Elementary–Age Children about Sexual Abuse**
>
> - No adult should touch a child's sexual parts except at a doctor's office.
> - Sexual abuse occurs when an older, stronger, or more powerful person looks at or touches a child's genitals for no legitimate reason.
> - A person who is sexually abusing a child may tell the child to keep the behavior secret.
> - Tell a parent right away if unwanted or uncomfortable touching happens.
> - A child is never at fault if an adult—even a family member—touches him/her in a way that is wrong or uncomfortable.
> - Most adults would never abuse children.
> - Both boys and girls can be sexually abused.

"But All the Kids Watch It"

The media—television, movies, even the news—are probably one of your children's most prominent sexuality educators. I have heard the Reverend Jesse Jackson say that television is now children's third parent. Television and movies teach your children about what it is to be a man or a woman, how men and women relate to each other, what is attractive, and even what is "sexy." And most of us don't like the messages that many television programs and movies give our children about these issues. Too often, television and movies are sexist, violent, exploitative, and degrading. Few programs model honest, equitable relationships between men and women, and almost none show people practicing responsible sexual behaviors.

Television shows contain an amazing amount of sexual content. In fact, you may be surprised to learn that according to a 2003 study by the Kaiser Family Foundation, about seven out of ten shows during the evening hours have sexual content, averaging about six sexual references per hour. Only about one of four shows that talk about or depict sexual intercourse also mention contraception and/or prevention of sexually transmitted diseases; but four years ago, only about one in six offer such messages. And a new study by the Rand Corporation showed that these messages make a difference in what young people know about safe sex: teens were able to recall a message in *Friends* about the need for proper condom use.

After sleeping and attending school, children ages six to eight spend more time each week watching television than they do playing, eating, doing chores, participating in sports, or attending church. American children watch an average of 24 hours of television a week. By the time an American child graduates from high school, they will have spent 18,000 hours in front of a TV set compared to only 13,000 hours in a classroom. And a child will view a staggering 20,000 commercials a year as part of that television time.

And few shows or movies are completely "safe." Even programs rated "G" or "family viewing" may have messages that are not consistent with your values about sexuality. Three out of four "family hour" programs contain some sexual content, up from 43 percent in 1976. Thirty percent feature scenes with a primary emphasis on sex, up from 9 percent in 1976.

Even children's movies aren't always safe. I remember taking Alyssa when she was about six to see a return of the animated classic *Peter Pan*. I was appalled by the helplessness of Wendy, the seductiveness of Tinkerbell, and the racist messages about Native Americans. Of course, I hadn't remembered any of that, and I had trusted Disney to offer "wholesome" fare. I could scarcely contain myself during the movie. Afterward in the car, I tried to talk with her about some of the images I found disturbing.

Think for a moment about these questions: Which television shows model adult sexual relationships in a way that is consistent with your values? Which ones don't? Which television shows portray boys, girls, men, and women in a way that you would like to teach your children? Which ones don't? And which shows do your children watch?

A surprisingly high number of children, particularly those in grades three to six, say that they are influenced by the content of television. In a study done by Girls, Inc., 51 percent of girls in this age group say they talk like a character they have seen on television, 37 percent say they have acted like a character on television, and one-quarter have worn their clothes or hair like a character on television. One in six say that they have dieted or exercised to look more like a character they have seen on television.

Short of removing the television from your home, what can you do? (I do have friends who are raising their children without televisions in the home. I think that's admirable, but I am much too much of a news junkie to do this. And to be honest, there are times when television and videos are a great way for all of us to unwind.) I suggest the following:

- Restrict the amount of television your child can watch. This not only makes sense regarding sexual issues, but it is also related to cutting down on the violence and aggression your child is exposed to. It also means more time for family walks, bike rides, puzzles, and talks. Consider putting your family on a "TV diet" and limiting television to no more than two hours a day. You may also want to develop a list of "approved" TV shows that your child can watch without an adult present.

- Use your own personal judgment in deciding which shows are appropriate for your child to watch. Don't trust the rating systems. Television shows rate themselves as TV-G, TV-PG, TV-14, or TV-M. In response to pressure, they also added more detailed ratings: S for sexual situations, D for sexual innuendo, L for vulgar language, and V for violence. The problem is that these ratings may not reflect *your* family values. For example, I am personally opposed to my child watching any violence, even cartoon-pretend violence. The morning cartoon shows may be rated G, but I'm not happy with Gregory's watching them. Watch all programs alone first and make up your mind before you let your child watch them alone.

- Don't trust the "family viewing hour." The so-called family viewing hour is between 8:00 P.M. and 9:00 P.M. during the week. Many shows with highly suggestive comment and innuendo appear then. Even if the shows are consistent with your values, the advertisements and promotions for the shows later in the evening may not be. Watch TV with your child or put in a trusty, prescreened video. This may be tough—family hour may seem like the best time to do house chores after a long day in the office or to have some private time after a long day with the kids. According to a study by the Kaiser Family Foundation, two-thirds of parents say they watch TV with their children a maximum of half of the time; their children report it is even less.

- Don't put a television set in your child's room. A surprisingly high number of parents of elementary-school children give their child a TV in their room. They say it is because they and their children like to watch different programs. Unfortunately, that may be exactly the point: Your child wants to watch programs that you may not want them to watch. And with the door closed, and a remote in his or her hand, you will never know. I personally don't allow my children to watch evening television without an adult in the room to discuss the issues that are raised; but at minimum, you want the TV in the den or living room where you can walk in and out and see that your TV rules are being followed.

- Do your research. Take the time to watch television shows before you give your child permission to watch them. Read the reviews of movies in the newspaper, especially what they say about applicability to children. A friend of mine, Nell Minow, is the Movie Mom and writes reviews each week of new movies playing in the theater and new releases on video and DVD. She rates movies on their suitability for children and includes good advice on follow-up discussions. Her reviews can be read at www.movies.yahoo.com/moviemom.

- Look for teachable moments. Say you are watching an 8:00 P.M. show and a situation arises that is not consistent with your family values. You can use this as a teachable moment. If you are really offended, you can turn the television off or change the channel, and explain that this program is off limits. And then explain *why*. It might be a better learning opportunity, however, if you wait for the commercial and then begin a dialogue with your child.

Here's how it might go:

Parent: It really upset me when that man yelled at his wife. What do you think?

or

> *Parent:* Remember when that woman put on the tiny, tight dress before her husband got home so she could get her own way? What might she have done instead?

Media and Values magazine had some advice for parents about using television shows to teach your child your family's values, and I've adapted them here:

- Comment on what's good and bad about the show's messages. Be sure to talk about what you do like as well as what you don't like. Ask your child for their opinion. Don't put down your child's taste in television shows; stick to the messages.
- Be selective about who's watching television with you when you bring these issues up. Your daughter's slumber party is probably not the time for you to dissect a show's sexual images. It may be embarrassing if her friends or relatives are there. Every show does not have to be an opportunity for a discussion about sexual issues.
- Keep discussions informal and fun. Use the commercial breaks for quick discussions. Don't lecture during the program.
- Don't worry if your child doesn't want to get into a major discussion with you. They are still hearing your reactions and opinions.
- Muse aloud. Sometimes it is more effective to present your reactions than to ask your child a direct question: "Wow, I can't believe that couple went to bed with each other on their first date."

You can also use these conversations as a way to introduce discussions of actions and consequences: "What is likely to happen next as a result of their behavior? What other choices might they have made? What might have happened then?" Use the situations set up by the show ("When Stephanie forgot to bring her homework to school, she lied to the teacher that she had lost it. What else might she have done?") to help your child understand that people need to think through the consequences of their actions.

In our home, we have set certain rules about television that seem to be working. First, the television is not allowed to be on between 9:00 A.M. and 5:00 P.M. This allows Gregory to watch one program on PBS in the morning, and Alyssa to watch one show before dinner. It means that soap operas and daytime talk shows are not coming into the house. Second, we watch evening shows with our children, and Alyssa knows that if they have any sexual themes (or, for that matter, other moral issues), we will talk about them. As she got older, this cut down on the number of shows she wanted to watch. Somehow, certain more racy shows are less fun when you are discussing the innuendo and sexism at every commercial! Third, certain shows are off limits. Until she was thirteen, Alyssa was not allowed to watch MTV or VH1; Gregory was not allowed to watch cartoon super heroes like Ninja Turtles or Power Rangers. We actually often choose reruns of sixties sitcoms on Nickelodeon, although even there, I feel it is important to talk about how men's and women's roles have changed since the 1960s. After all, how come Lucy, Laura Petrie, and Samantha Stevens never had to go to work?

You should also know that even with your best monitoring, your child may watch some shows that you would prefer them not to see. They may see them at a friend's house. Your baby-sitter may "forget" your rules. They may actively seek to watch "forbidden" programs when they are staying over at your in-laws'. On more than one occasion, Gregory has told me about the super heroes he watched at a friend's home. This type of occasional viewing is certainly not going to hurt your child; if you discover it, it may even give you the opportunity to reinforce your family values. And sometimes, it's not the entertainment shows that teach your children about sexuality:

Conversation Starters for Television, Ads, and Movies
- Gee, that teenage girl looks really skinny. I wonder whether you think that's attractive.
- I hated that commercial that showed that using the right shampoo means that boys will like you.
- I liked that the couple talked about whether they should have sex instead of just doing it.
- It's amazing how many families on TV have only one parent.
- Which of these characters would you like to have as a friend?
- Which of the adult characters would you like to have as a parent?
- Which programs that you like show women and girls the way you want them to be pictured?
- Which programs that you like show men and boys that way?

Some Important Messages about the Media for Five- to Eight-Year-Olds
- Some of the material on television, in the movies, in books, and on the radio is true and some is not.
- Some media make people and things look different and better than they really are.
- Some media are not appropriate for children your age.

Lately, even news programs aren't safe for your child to watch alone.

The News

It is hard to believe how often the news addresses a sexual issue. In 1999 we had a national teachable moment when President Clinton faced impeachment for lying about his sexual relationship with an intern. That news was on the front page of every paper and the focus of every news show for almost a year.

It is not unusual for sexually themed news stories to enter your home or your child's consciousness. Think about some of the news stories of the last few years: priests and sexual abuse, gay marriage, sexual harassment in the workplace, new discoveries on AIDS. In fact, every day there are stories that deal with sexuality in the press. And it's not unexpected for your child to be exposed to them. I remember when Alyssa was in the first grade, she came home from school and asked, "Mom, who is Magic Johnson and why is everyone talking about him?" (Magic Johnson had just disclosed his HIV status to the public.) Her friends in the second grade were also obsessed with the stories following the Polly Klaas murder, something that I had not brought up for discussion in our home until I found out she knew about it. If it is happening in the news, your elementary-school child has probably heard somebody talking about it.

Even younger children absorb what's going on around them. In 1999, we were on a long family car trip, and we were playing "20 Questions" to help pass the time. Well, we had determined that it was a living woman who had been in the news. Four-year-old Gregory piped up, "Is it Monica Lewinsky?" Stunned, I turned to the back seat, and said, "Who's Monica Lewinsky, Gregory?" He answered, "Isn't she the teenage figure skater?" "No," I answered, struggling not to laugh, "that's Tara Lipinski!" But the point is that somehow, he had picked up Monica Lewinsky's name from overhearing news, radio, or adult discussions.

These news stories present a unique teachable moment. And if the story is big enough, like the charges against President Clinton, I don't think you can ignore talking about it with your child. A colleague of mine called me during the Clinton intern flap: His six-year-old daughter greeted him at the door, "Dad, do you know that Bill Clinton is dating?" But it was the nine-year-old's statement that really threw him, "I heard he likes people to kiss his private parts."

By this point in this book, you know that ignoring issues doesn't help your child. It only gives them the opportunity to get the information or misinformation from another source, and it teaches them that in your house, parents don't talk about sexuality issues. But how do you start?

The first thing you always want to ask yourself is, "Which messages do I want to get across about this issue?" If the story is about same-sex marriage, then this could be your chance to share your family's values about homosexuality as well as employment or housing discrimination on the basis of sexual orientation. If the story is about extramarital affairs, this is an opportunity to give your values about fidelity. With a preadolescent child, it is a chance to talk to them about your hopes and expectations of their own behaviors: "Sexual feelings can be very intense, but you can have sexual feelings without acting upon them, especially if it might lead to harmful consequences for yourself or your partner."

Next, ask yourself, "What is the easiest time to bring this up?" Are you comfortable having this as a dinner table discussion: "Did anyone see that story in the news today about the new treatment for AIDS?" Or would you rather have this as a private discussion after school or before bedtime, "Honey, I wonder if you heard anything in school today about the woman who had seven babies?"

Start by asking them, "What did you hear?" followed by "What do you think?" Give them time to tell you what they already know, and be sure to correct any misinformation. Then share your values about this issue with them.

It's not just the news or television that may introduce your children to values different than the ones you hold dear. Sometimes it's the school.

Working with the Elementary School

One of the challenges for parents when their child enters kindergarten is developing a partnership with the school. Parent involvement in schools is critical, and parents support their children's success in school in many ways. Children who do well in school have parents who

- talk to them about what happens in school.
- have high expectations for their children's success.
- have close relationships with their children.
- stay involved with their children's lives.
- believe in their children.
- monitor what is happening in school, including being sure to attend teacher conferences, volunteering for field trips, and asking to observe in the classroom.

What does all of this have to do with sexuality issues? Building self-esteem in early elementary school has everything to do with self-esteem in later childhood and even adulthood.

Children who feel good about themselves are more likely to be successful and happy. They are less likely to be swayed by peer pressure and less likely to engage in risky behaviors, including drinking and unprotected intercourse, as they grow up. Adults with higher self-esteem make better decisions in their lives, including love, romantic, and sexual decisions. Children need to learn that they are loveable and capable. The schools play an important role in helping you as a parent instill self-esteem in your children.

I would like to think that all schools are concerned about encouraging self-esteem in children, but that is not always the case. When Alyssa was in kindergarten, we had a very disturbing situation at her first science fair. The kindergarten children were invited to do science fair projects (in retrospect, a somewhat dubious assignment at best). Alyssa had done a very sweet project on our cat's new kittens: They had been born at the beginning of the project time, and she had measured each of them each week to monitor and compare their growth. The project, called "My Cat Had Kittens," included photos she had taken herself with an instant camera. The night of the science fair, she proudly took us into the gym to see her and her classmates' projects—and her pride crumbled when she saw that three of her kindergarten classmates had "won" ribbons in the science fair and she had not. She left sobbing, "Why wasn't my project good enough?"

We were livid. It had never occurred to us that the school would be judging kindergarten science fair projects and that some children's projects would be deemed better than others. When I called the principal the next morning, he indignantly replied, "What did you do to push her to win?" I replied, "We didn't. We did not even know that there was a contest!" I tried to talk to him about how detrimental this had been to her self-esteem; surely, all the school had wanted to do was have these five-year-olds feel proud that they had completed a science fair project. I ended up having to go to the director of science for the city's schools with my concerns; and yes, there is now no judging of any science fair project until the child is in the fourth grade in our city!

The point of sharing this story with you is that you are your child's best advocate at school. You need to assure yourself that the school and the teachers are committed to helping your child develop a sense of their self-confidence and their capability. And you need to stay involved throughout your child's years in school.

There may be many times during your child's elementary-school years when the school teaches values that are different than your own. This doesn't only happen in health classes; it is just as likely in science, reading, and social studies lessons. For example, what if your child is assigned a book to read that you don't like? Or, in our case, how did we deal with the school's talking about Indians and portraying Columbus as a hero, when we had told our children that the correct term is "Native Americans" and that Columbus didn't "discover" America? Sometimes you need to just make sure that your child knows that your family's values are different, and sometimes, you may want to intervene.

My experience with my children's schools is that you have to pick those times carefully. You don't want to become known as that interfering parent who is in the principal's office each week. On the other hand, you do want the school to know that you are in partnership with them and that you will advocate for your child.

One issue that I have consistently chosen to address is how women are portrayed in the curricula. Too often, even today, women are invisible and anonymous in history, literature, science, and social studies lessons. For example, when Alyssa was in the fourth grade, she was in a class that was studying the pioneers. Each child was assigned a family to follow as they crossed the country. Alyssa came home concerned, "Mom, my family is called John Smith, his wife, and his son Bobby. How come his wife doesn't get a name?" I first suggested to Alyssa that she ask the teacher if they could give all the women first names at the start of the next class. And then I called the teacher to see how the contributions of the women pioneers would be addressed in the curriculum. She said—to my surprise—that she had never really thought about it before. I asked her to make sure that there was information presented about Calamity Jane and Annie Oakley, as well as that each lesson discuss the contributions of both the pioneer men and women. Making sure that both elementary-age boy and girl children learn about competent, important men's and women's accomplishments lets them know that both men and women contribute to society—and that they can grow up to make important contributions.

The playground and the school bus are also fertile grounds for your child to learn about sexuality issues.

"Tag, You Have AIDS"

A few years ago, I received a call from a concerned first-grade teacher. During recess, her students were playing "Tag, you have AIDS" and then dissolving into laughter. She didn't know what to do.

Think for a moment. Children and teens today have always had AIDS in their world. The first cases of AIDS were diagnosed in 1981; the virus that causes AIDS, HIV, was discovered in 1985. Unlike those of us who had to learn about AIDS and

HIV, today's young people have always had HIV and AIDS as part of their environment, in the same way strep throats and tonsillectomies were part of our childhoods.

Your elementary school–age child hears about AIDS—in the news, on television, and perhaps on the bus and play-ground. They need to know certain very basic information.

First, this is an opportunity to teach them some basic good hygiene and health practices. AIDS is caused by the human immunodeficiency virus (HIV), just as everyday colds are caused by a virus. Washing your hands before eating or after being on the playground, covering your nose when you sneeze, and not rubbing your eyes are good ways to reduce the chance of getting colds. Not getting other people's blood in your body can keep you safe from HIV: Children should be told that they are never to touch anyone else's blood, even to help a friend with a bloody nose, or to become "blood brothers." They should know that they should seek a grown-up if they come upon needles in the schoolyard, playground, or street, or if a friend cuts herself.

Second, it is very important for children to know that they do not have to worry about getting AIDS from just being around people who have AIDS. Unfortunately, there are still too many myths about HIV and its transmission. The only ways to transmit HIV are through sharing needles, unprotected sex-ual intercourse of any kind, and infected blood products, and during pregnancy from a pregnant woman to her fetus or dur-ing breast-feeding. Mosquitoes, visits to the dentists, and food *do not* transmit HIV.

Both you and your child need to know that there is no rea-son to fear a child who is infected with HIV in their classroom. There have been some horrible cases of discrimination against children with HIV in schools; generally, they have arisen out of people's fears that their children will be exposed to the virus. Schools have strict procedures for handling the possibility of a child with HIV in the classroom, including how to deal with blood spills. Children cannot get HIV from another child simply by playing with them, sharing food with them, sitting with them, or talking with them.

This can also be an opportunity to talk with elementary school–age children about compassion. Children this age can understand that AIDS can happen to anyone who practices certain unsafe behaviors, and that people with HIV and AIDS deserve our support. If you know people who are HIV positive, your children can benefit from knowing this about them and learning that they shouldn't be afraid.

"Oh S@#$!"

One of the other things your child may be learning on the elementary-school playground is inappropriate, "bad" language. It is very likely that one day, when you least expect it, your five- or six-year-old is going to use the "S" word, the "D" word, or even the "F" word.

Before you explode (or burst into laughter), first think about why they might be using these words. They may be trying to get your attention. They may be trying to shock you with their new, grown-up vocabulary. They may even be innocently repeating words that they have heard you use!

Next, think about the messages you want them to learn about this kind of "dirty" language. Also, think about whether you and your partner are using these words at home. What power do these words have for you?

I have to admit that I went to college at a time when the "F" word punctuated nearly every sentence, and I couldn't get too worked up when my children tried out these words. But it was important to teach them that *other* people might get very upset if they used that language, especially other adults.

Your reaction will have a lot to do with whether your child continues to swear. If you laugh, they will think it's cute and continue to use these words. If you scream or punish them, you are likely to drive the behavior outside of your home but not eliminate it when you are not around. You may also give the words more power, which will mean their use can become part of a power struggle between you and your child.

In our home, we taught our children that certain words evoke very strong reactions and are not "nice." (And we include "shut up" and "hate" in that list.) I asked them not to use those words and said that we wouldn't use them either. I've even heard of couples who fine themselves one dollar every time they swear in front of the kids to break themselves of the habit.

You can also use this as an opportunity to talk to your child about their feelings. If your children swear when they are angry or mad, try to label the feeling: "You must be feeling pretty angry that you dropped your milk and now you have to go back and clean it up. What other words can you use when you are mad that aren't swear words?"

By the way, it probably won't work to use this as a moment to discuss the meaning of these words. I tried to do this once, with very little success. Eight-year-old Alyssa and I had gone for a walk to a park, and on the bench we were sitting on, someone had written the "F" word. She said, "Ooh, mom, look at that word." I thought, "teachable moment!" I answered, "Isn't it sad that this ugly word describes how adults [notice adults] love each other." She looked at me with disgust, "Mom, do you always have to be a sex educator?" And then she walked to the swings. See, even I don't always get this right. Seriously, I would not recommend using these opportunities to talk about defecation, damnation, or intercourse—look for other ones.

Your children may also ask you about sexual behaviors that they have heard about using this type of language. Five-year-old Sheila asked her mother during the car pool ride home, "Mom, what's a blow job?" Six-year-old David asked his father, "Dad, do you and mom f***?" This type of language does seem quite shocking coming out of the mouth of your innocent five- or six-year-old. Try to remember that your child is most likely repeating something s/he has heard on the playground or on the school bus from an older child. Calmly ask them, "Where did you hear about that?" followed by "What do you think that means?" Then give them a short, simple answer, which will probably satisfy most children their age. You also will probably want to ask them not to use those words again.

> *Parent*: That's a not very nice term for adults making love. It's a word that most adults don't like, and I hope you won't use it again.

"The Britney Syndrome"

I attended the Halloween parade at Gregory's elementary school every year. In every class, there were police officers, ogres with scary masks, and a few children dressed in tie-dye outfits, going as 1960s hippies in the same way that I once wore a poodle skirt on Halloween. And in almost every class, from kindergarten through fifth grade, there was a girl in a sparkly midriff top, a bare stomach, and low-slung tight matching pants, and in some cases, thigh boots. When I asked one who she was dressed as, she answered, "Britney Spears."

Now, this will come as no surprise to parents with daughters in elementary school. What I call "The Britney Syndrome" has been around for years, even though Britney herself is now trying to appeal to an adult audience. Britney Spears has gone from proclaiming her virginity—despite songs like "...Baby One More Time"—to kissing Madonna open-mouthed on national television and appearing nearly nude in *Esquire* magazine.

Nevertheless, Britney's appeal (and Lindsay Lohan's and Beyoncé Knowles's) is still largely among very young girls, many of who seek to mimic their dress and sashaying style. Pop stars are not alone in selling sex to six- to eight-year-olds. The biggest doll sensation in recent years was the Bratz dolls. Here's how one Web-based advertisement describes them, "the Bratz have it all: looks, cars, makeup, and the latest clothes. With a hip line of dolls and games, the Bratz offer girls a world filled with funky makeovers, endless shopping, and lots of time for just plain kickin' it in style." The dolls are dressed in midriff tops, bare stomachs, low-slung pants, and platform sandals. Indeed, the overall look, according to Ruth La Ferla in the *New York Times*, "might be at home on any street corner where prostitutes ply their trade." Although the Bratz Web site says that the dolls are aimed at an eight- to twelve-year-old audience, my son Gregory tells me that they are mostly popular with the first-through third-grade girls. Barbie, that fashion icon doll of more than forty years, is most popular now with three- to five-year-old girls.

At almost every talk I give at an elementary school, a parent asks me what to do about their daughter's desire to wear low-slung jeans and skimpy midriff tops. Parents generally indicate that they are uncomfortable buying or allowing their daughters to wear these clothes, but that they also somehow feel badly about it. At a talk, one mother actually said this about her eight-year-old, "I don't want to inhibit her fashion sense."

Shopping for clothes has become another teachable moment. Ask your daughter what messages she wants to give her friends and the adults in her life with her clothes. Explain to her that the adults may see a young girl in a midriff top and tight pants as wearing "sexy clothes."

Yes, you can use the word "sexy." It is not unusual to hear comments about being sexy in your home. It is important to talk about what sexy means. And to emphasize that sexy means different things to different people. First, find out what your child thinks sexy means. You might be surprised by their answer. One friend reports that her child said, "Sexy is taking off your clothes together!" Another said, "Isn't that when you make babies?"

As always, try to correct misinformation and then give your values. You might say, "Sexy is a way that adults describe someone they find attractive. I think Daddy looks sexy when he wears his jeans and black shirt. I don't like you describing yourself this way because it really is about adults."

And then see where the conversation goes.

You *can* set limits for your children's clothes. It can be hard to find appropriate play and school clothes for your daughter as stores have kept up with these fashions. If you can't find what you are looking for, get catalogs from places like Land's End or L.L.Bean and give your daughter the freedom to pick out whatever she wants from one of these catalogs. If you feel your daughter simply must have one "Britney-like" outfit, let her know that she can't wear it to school, church or synagogue, or her relatives' homes.

You also can set limits on which dolls are appropriate for

your child. American Girl dolls offer young people a chance to learn about the nation's history and women's role in it. In the words of one ten-year-old girl, writing a letter to the *New York Times*, "Bratz dolls sometimes leave a bad impression. When girls get older, some will probably wear the same clothes that these dolls wear. My dad says the makers of these dolls care too much about their looks. It seems that the only taste they have is for shopping and fashion, not reading, writing, or thinking…here's my point: It doesn't matter what you see on the outside, but what is inside your heart and head."

The dolls and the clothes also are teachable moments to give your elementary-school child some quick messages about body image. Our cultural messages about thinness and beauty reach children early and set the stage for future unhealthy behaviors. Talk to your child about the fact that Barbie, Britney, and the Bratz give unrealistic images of women's bodies—a message that is important for your son as well. Be sure your child is eating nutritious foods and is not overly concerned with her body. In one study of elementary-school children, more than 40 percent of fourth-grade girls said that they had been on a diet.

Elementary-school children need to know that bodies come in different sizes, shapes, and colors, and that all bodies are special. They also need to know that male and female bodies are equally special. And they need to be encouraged to eat well and exercise. Fortunately, today's children are being exposed to a wide range of physical activity earlier than we were; in my town, it is not unusual to have five-year-olds playing soccer or taking karate lessons. And although I am personally against the high-pressure aspects of some childhood team sports, I think teaching every child that their body is good and capable is an important foundation for adult physical and mental health, including sexual health. Every child should know that they can be proud of the special qualities of their body, regardless of size, shape, or physical attributes.

Special Issues

Divorce

It goes without saying that divorce is difficult for children. More than one million children go through a divorce each year. Helping a child cope with divorce is well beyond the scope of this book; there are some excellent resources that can help you.

Diane Papilia and Sally Wendkos Olds have identified a set of guidelines for dealing with divorce and children.

They suggest that

- all children in the family be told at the same time;
- children be made to understand that they did not cause the divorce and they cannot change the parents' minds;
- custody arrangements be explained in detail—and explained again;
- children be told that both parents still love them;
- children be encouraged to express their feelings about the divorce—not only with their parents but with other caring adults;
- parents never put children in the middle of their disagreements;
- children's lives be changed as little as possible; and
- counseling be made available.

Being a child of divorce does seem to affect children's sexual development. Girls with divorced parents are more likely to become involved earlier with dating and sexual relationships. In contrast, teen boys with divorced parents are less likely to initiate sexual relationships. And close to two-thirds of teenagers with divorced parents believe that there is a high possibility of divorce in their own future marriages; girls are especially worried that they will not form lifelong relationships.

And Olds suggests that "parents must recognize the real conflict between their needs and their children's needs." And this is the heart of one of the sexuality issues raised by divorce: how to fulfill your adult needs for intimacy and relationships and still put your children's needs first.

Parents Dating

You may have had your child as a single parent or you may be recently divorced. Regardless of your situation, your elementary school–age child is unlikely to be happy about your dating.

Remember *The Parent Trap?* In both the original movie with Hayley Mills and the 1990s remake, the twins scheme to get their parents back together. Children of divorce often fantasize about their parents getting back together, and it is difficult for children of all ages to first accept that their parents will be emotionally and physically intimate with another partner. I was thirty-three when my mother remarried. And I was surprised and amused at myself when this thought popped into my mind right before the wedding, "Well, I guess Mom and Dad really are not going to get together!" And they had been divorced for more than a decade!

You need to think carefully about how you will deal with dating. If you are divorced and have a shared custody arrangement, you may want to keep your date nights to the nights that your child is with your ex-spouse. You will want to tell your child that you are starting to meet and go out with new people, but that your time with her or him is still the most important time for you.

I think it is very confusing to children to be introduced to a string of mom and dad's dates. I also feel very strongly that it is not appropriate to have children wake up to find a stranger in their parent's bedroom or in a robe in the kitchen— especially if parents don't want this to happen to them in a few years when their child begins dating!

My own bias is that only when you are becoming seriously involved with someone is it time to introduce your new partner to your child. Start out with casual, short encounters, maybe in public places. Introduce your new significant partner slowly.

Some of my single friends who are parents disagree with this advice. They tell me that they want their children to meet the people they are dating, even if it isn't a serious relationship. They say that their children can learn about dating relationships from watching them date different partners. They point out that it is healthy for children to understand that parents have needs for adult companionship and friendship. They also say that they don't like keeping secrets from their children.

You will need to decide what is right for you and your children. Be sure, though, to remember that your dating behavior sets a model for your child's future dating.

Sex at home when your children are there is a different matter. If you want to have your partner start sharing your bed in your home when your children are around, be sure to talk with them about it first. Do not let them discover you! Be sure to frame this in a way that is consistent with the values you want them to adopt for their behavior when they begin dating:

> *Parent*: Susie and I have been going out for six months now. We love each other a great deal. When grown-ups love each other, they start to want to be with each other all of the time and have time in private to show their caring. I'm thinking that I would like Susie to stay over in my room. Tell me how you feel about that.

What have you done with this simple exchange? You've told them that (1) sex and love go together; (2) sex is an adult behavior; and (3) you care about their feelings. Think through the messages you want to give them about their dating when they are older, and be consistent with your own behavior.

When Your Child Isn't Conceived the "When Mommy and Daddy Love Each Other" Way

The question "Where did I come from?" is much more complex to answer honestly for an increasing number of couples. Each year, more than 120,000 children are adopted and more than one million children live with adoptive parents; more than half a million children have been conceived through artificial insemination; an estimated 3,500 babies were born in 2000 alone from donated eggs; and increasing numbers of gay and lesbian couples are adopting or having children through assisted reproductive technologies. The number of live births from assisted reproductive technologies each year has increased 73 percent since 1996! In fact, with today's new technologies, a child could actually have as many as five people who participated in their conception and generation: the mother and father who raise the child, the woman who donates the egg, the man who donates the sperm, and the surrogate mother who carries the fetus to term. The answer "When mommies and daddies love each other a baby begins" certainly doesn't cut it in these situations or in many others.

According to the National Council for Adoption, most professionals believe that adopted children should be told that they are adopted starting at around the age that they ask where babies come from. Some parents decide to talk about adoption from the time they bring the child home. They figure that by having adoption be part of continuous conversation, they will never have to reveal this as a family secret. Other adoptive parents think that between the ages of three and five is the best time to first reveal this information; they wait for children to ask the dreaded "Where did I come from" question. Some professionals recommend that parents should wait until age five: They believe that children must have reached a certain maturity before they can handle this type of information.

Joining an adoptive parents group where you and your child can meet other families with adopted children can be very helpful. It lets your child know that adoption is common and normal, and it gives you other parents to talk to about the special challenges of having an adopted child. Look in the appendix for places to go for more information.

Preschoolers and even early-elementary-school children won't understand the details of infertility or adoption, but they can be helped to understand, in psychologist Anne Bernstein's words, "the special story of how you came to us." They can be told that another mom is their "birth mother" but that you are their real parents who will love them forever and raise them to adulthood. I like this explanation, adapted from *How to Raise an Adopted Child*, to the preschool question "Where did I come from?":

Parent: You grew in the uterus of another mommy, who

could not take care of you. We wanted a baby very much, so we took you home and love you very much.

That should satisfy most preschoolers and early-elementary-school children. Experts in adoption say that children will not fully understand the concept of adoption until much later: They may be in middle adolescence before they completely integrate the idea that they were adopted into their being. Talking with your child about their adoption is an ongoing process. Books and videos can help. See the appendix for information that can help you.

Elementary-age children are likely to ask, "Why did my mommy give me up?" and may mistakenly believe that they did something to make this happen. It is important that you reassure them if they say, "It's my fault that I'm adopted," and try to be honest about the reasons, if you know them, why their birth mother gave them up for adoption.

> *Parent*: Your mother was very young (poor, sick) and wanted you to have a better life than she could give you. She cared about you very much, but before you were born, she decided that she couldn't be a good parent to you.

Bill Pierce, president of the National Council on Adoption, cautions that you should not tell your child that they were placed for adoption because "your mother loved you very much." One problem with this is that children may think that you will place them up for adoption if, for example, your family begins to have financial problems. Second, you are unlikely to know how the birth parents really felt about your child. You can say something like:

Parent: I don't know your birth parents, but I do know that they cared about your well-being and happiness and wanted you to find parents who could care for you and love you like I/we do.

I have been told that many adopted children, after learning the facts of reproduction, tell their mother, "I wish I could have grown in your uterus," expressing a need for this type of closeness. It is important to acknowledge this feeling. You could say something like:

Parent: I understand how you feel that way. I wish that too. But my body couldn't have a baby, and I am so lucky that I have you. Do you want to hear the story about your adoption day?

For more information about talking to your adopted child about their "special story," you might want to contact the adoption resource organizations listed in the appendix.

Children conceived by in vitro fertilization raise other issues. Some professionals believe that when the parents are both the sperm and egg donor, as in most cases of in vitro fertilization, children do not need to know the specifics of their conception. Other parents want to share the unique details of their child's conception. It is up to you.

Parents of children who were conceived with artificial insemination by a sperm donor or with donated eggs have a more difficult problem. In most cases, these parents do not know the identity of the sperm or egg donor. (But they are likely to know the healthy history of the donor, information that at least will need to be shared with their teenage or adult children who need to know how to arrange for their own medical care.)

Professionals are currently debating whether children should be told that they were conceived using sperm or eggs from an outside donor. After all, unlike adoption, the mother carried the fetus to term and delivered the baby. When the father is not the source of the sperm, some professionals believe that because of health histories, it is critical that children learn this information before they need it in the future when they will be taking care of their own health needs. They also argue that these types of family secrets can be devastating to the family and to the child if they are discovered by accident. Other professionals believe that this is a highly individual decision and each family needs to decide for itself. They cite fear of social disapproval and criticism as well as the need to protect the family's privacy—after all, a four-year-old could indiscriminately share this information with others.

This is a highly individual family decision, and one that you should discuss with your partner and trusted counselors.

Women who are single mothers by choice or lesbian couples or men who are adoptive gay fathers still face other issues in answering "Where did I come from?" Until children are in elementary school, they are likely to believe that their own experience is universal: All children have two mommies, or two daddies, or no daddy or mommy in their household. It is only when they begin interacting with other children in nuclear family relationships that they begin to question their own situation. "Do I have a daddy/mommy?" is a question that can be expected by the age of three or four from a child of same-sex parents.

Psychologist Cherie Pies suggests that lesbian mothers think through a variety of questions about their children's father. I've adapted some of these questions for gay men who are fathers as well.

- How and when do you want to tell your children about their biological mother/father?
- How will you answer "Who is my mommy/daddy?"
- How will you answer "Why don't I have a mommy/daddy?"

Your answers may differ if the child has been adopted or created by artificial insemination. It makes a difference if you know the woman who carried your child or the man who fathered him/her, or if you will, as in the case of lesbians using a sperm bank, never know them. In the case of artificial insemination, you could say, "I/We wanted a baby very much. So I asked Uncle Ben to help me. He provided the sperm that joined with my egg to help get you started inside me." In the case of using an anonymous donor, you could say something like, "I/we wanted a baby very much. We went to a sperm bank to get the sperm that started you. Let me tell you what I know about the man who helped get you started."

Perhaps the most important thing you can do for preschool- and elementary school–age children is validate that there are many different kinds of families. There are many books on this for small children; there is even a Barney song celebrating diversity in family lifestyles.

In elementary school, your child may become aware for the first time that your family is different than other families. It is important that you prepare them, and that you discuss with them how you want them to share information about your family. One of my friends worried that her five-year-old child would proudly announce to the playground, "My mommies are lesbians," and that it may not be heard kindly by his classmates or other adults. Your child could be teased by less tolerant classmates. These are important issues to discuss with your child and your child's teachers.

Like heterosexual adopting parents, gay and lesbian parents may benefit from meeting with families that are similar to them. In some communities, this is quite easy as there are gay and lesbian community centers that hold parenting groups. There are estimated to be more than 200 gay parenting groups across the country. In other places, it may be more difficult. There are now an estimated six to ten million lesbian and gay parents in the United States. There are resources and newsletters that may help you: try *Gay Parent* magazine, www.gayparentmag.com, or *Love Makes A Family*, www.lmfamily.org. There are also some books listed in the appendix.

Values Exercise for Chapter 5

Your eleven-year-old daughter or son has been invited to their first boy/girl party. You have heard from other parents that kissing games like "spin the bottle" are common at these parties. You...

❏ a) tell her or him that s/he's too young for boy/girl parties.

❏ b) tell her/him to go and have a good time.

❏ c) call the parents of the child who is hosting the party, talk to them about their plans for supervision, and then decide.

❏ d) find out what your child's friends are doing.

You walk into your twelve-year-old's room with a stack of clean laundry and discover him/her on his/her bed masturbating. You...

❏ a) walk out and say nothing.

❏ b) say, "Good for you. I'm glad you've discovered self-pleasuring."

❏ c) yell, "Stop that right now."

❏ d) say, "I'm sorry, I should have knocked," and leave.

Your son/daughter wants to dye his/her hair purple. You...

❏ a) say, "Not in my house, you don't."

❏ b) say, "Tell me more about why you want to dye your hair."

❏ c) shrug and let her/him do it. After all, hair dye is only temporary.

❏ d) ask them, "What message do you want your appearance to send?"

You are putting away a book in your son/daughter's room and you find a note in his/her handwriting on their desk, "I Love Chris," with a big heart. You...

❏ a) ignore it; you probably weren't meant to see it anyway.

❏ b) ask as soon as you pick your child up at school, "Who is Chris?"

❏ c) try to listen in on your child's conversations to see if you can find out more.

❏ d) at bedtime say, "I saw your note about Chris today when I was putting a book away in your room. Tell me more about Chris."

Chapter 5
Upper Elementary School
Ages 9 to 12

Your nine- to twelve-year-old is turning into an adolescent. They are going through the most rapid period of physical, social, and emotional development since infancy and toddlerhood. However, while parents are often well prepared for the developmental changes and milestones that occur from birth to two, they are often left in the dark about what changes to expect during preadolescence and adolescence. Sure, as parents we know that during these years, our children will get taller and go through puberty, but we are often not prepared for the social and emotional changes. It may help you to know that most of these are as predictable as the developmental milestones of the first year of life. As Alyssa went through these years, knowing what to expect made some of these challenges a little easier to live through—for both of us!

Puberty, *teenager*, and *adolescence* are not interchangeable terms. Puberty is the stage of maturation when a human being becomes capable of sexual reproduction. Teenager is defined chronologically: A teenager is a young person between the ages of thirteen and nineteen. Adolescence is actually a relatively new concept: It is the period of development that extends from puberty (or the end of childhood) through the attainment of adulthood. Prior to 1900, and still today in many developing societies, children marry shortly after puberty and begin their adult responsibilities. Today, as more and more American young people stretch their college and graduate school education through their twenties and even thirties and then return to live in their parents' homes, some have wondered when contemporary U.S. adolescence ends—at 30? when you buy your own home? have your own children? before your mid-life crisis?

Preparing for Puberty

Young people, some time between the ages of eight and sixteen, go through a predictable process of biological development called puberty. Normal pubertal changes may begin as early as age eight for girls and as early as age nine for boys, but some teens may not begin these changes until they are fifteen or sixteen. The average age for puberty to begin for boys is between eleven and twelve years; for girls, it is between ten and eleven. Still, "average" means that half of young people begin this process earlier, and half will begin it later. The process of puberty from the first physical changes to obtaining a fully adult body may take four or five years. On average, boys begin and end puberty about a year to two years later than girls do.

The first noticeable sign of puberty in girls is generally the development of "breast buds"—the breasts begin to elevate as small mounds. Later, the breasts and the nipples will get bigger. A 1996 study of more than 17,000 girls found that on average African American girls start developing breasts just before age nine and white girls start at about the age of ten. Light, sparse pubic and underarm hair begin to appear about six months later. Girls will begin to experience a growth spurt, often growing as much as several inches in a year. Their genital and underarm areas begin to produce sweat glands, and body odor may result. Most girls will begin menstrual periods about two years after breast budding begins, but some girls have completed breast development before their first period, and some girls' breasts continue to develop for many years after.

Boys experience many of the same changes. During puberty, they too develop light, sparse pubic and underarm hair, and their sweat glands become activated. Between the ages of twelve and fourteen, their penis and scrotum begin to enlarge. At first, the testes become larger, and the skin on the scrotum reddens and coarsens. As boys mature, their penis starts to grow longer. Pubic hair slowly becomes darker, and it starts to curl.

Girls today are experiencing first menstruation at earlier ages than when we were growing up three or four decades ago. The average age for first menstruation in the United States was twelve and a half in 2004. That means that half of girls will have their first period before the end of seventh grade, and many will begin as early as fourth or fifth grade. It also means that about half of girls will be in at least the eighth grade before their first period, and some may be as old as juniors in high school before they get a period. All of these are normal.

Young people today are reaching sexual maturity at much younger ages than they did in the past. Records from family bibles around the time of the American Revolution indicate that girls on average were seventeen when they got their first period. In 1860, the average age of a girl when she got her first period was somewhat more than fifteen years in Europe and North America. During the past century, largely as a result of improved nutrition, the average age at first period has declined an average of three months per decade, although it has remained fairly constant since 1960.

Preparing Your Daughter for Puberty

Your daughter is likely to notice that her body is changing before you do. She sees the stray hairs under her arms; she sees the few hairs peeking out on her vulva. She may notice a new smell under her arms or a new stickiness in her underpants. She may experience "growing pains" in her chest.

Many parents aren't privy to these changes. In fact, many parents are surprised when one day they look over at their ten-year-old daughter and realize that seemingly overnight, their daughter looks like she has sprouted breasts.

In most girls, what doctors call "breast budding" occurs at around ages nine or ten, but can begin at early as seven and as late as thirteen and a half. During puberty, the breasts develop, and the internal (ovaries, uterus, vagina) and external (labia, clitoris) sexual organs increase in size. The uterus actually increases in size by five to seven times. On average, the process of puberty in girls takes four years, but it can be as quick as a year and a half to as long as eight years.

In 1996, a study appeared in the medical journal *Pediatrics* that surprised many professionals and parents. It found that at least one-quarter of girls are beginning puberty, as marked by the appearance of breast buds, by the third grade. And yet, many parents wait until fifth or sixth grade to prepare their daughters for their first period. The age of first period, known to health-care professionals as "menarche" (pronounced men-ar-key), depends on numerous factors: race, genes, nutrition, and culture. And for unknown reasons, it occurs later in rural communities and in big families.

Too many young women say they are not being prepared adequately for these pubertal changes. Some girls receive only minimal education in school or at youth group meetings. I still vividly remember the Disney movie I saw thirty-five years ago in a fifth-grade Girl Scouts meeting. I mostly remember that we were all very embarrassed and quite nervous about this change that was coming. Things may not have changed much. In a recent focus group on adolescent female health, one twelve-year-old girl said:

> That was a big thing in the fifth grade. We had boys' education and girls' education. There was this little pamphlet we got. My friends and I were upset for the whole week.

Other girls report that they received only scant information from their mothers or older siblings or friends. In the same focus group, one girl reported, "On my tenth birthday, they gave us this little package and it was like a pad and a little booklet." Another girl reports that she really didn't get the information she needed: "I think that if I knew what it was, that all girls get it, and I was told that this is going to happen to everyone, that it doesn't hurt, and that it's fine, I think it would have been all right. But in a sense, you feel such shame."

Clearly, you want to do more to prepare your daughter for her first period. I am not suggesting however that you do this in a Big Talk. Remember, Big Talks are likely to be uncomfortable for everyone—you and your daughter. You want to prepare your daughter for puberty the same way you have handled these other situations—little by little, always being available for more questions. There are lots of way to introduce this: You can talk to her about your first period one day; introduce sanitary protection on a trip to the store when you have your period; tell her that you have noticed that her breasts are beginning to develop and that this is the beginning of wonderful changes in her body; give her a special present of a book on puberty; participate in a mother/daughter workshop. Some of these ideas are discussed in more detail below.

It helps if periods have always been a part of your household. When small children are introduced to this idea, it's easier to begin to talk around the age of seven or eight about the changes your daughter will be going through. My practice has been to keep my pads and tampons in the bathroom during my period and to answer questions about them when they arise. I've tried to give a message to my children right from the beginning that periods are a fact of life for women and that they are easily managed.

You might want to ask your pediatrician for advice about talking to your daughter about puberty at your child's annual physical. Your pediatrician can tell you how developed your daughter is. Many pediatricians use something called a Tanner scale to rate physical sexual development in both boys and girls. (See the boxes on pages 153 and 158.) A girl's first period is likely to come about two years after breast buds first develop, although this is not always the case. However, once your daughter has breast buds, make sure you have had several discussions with her about what to expect when she gets her period. It is a good idea to have her keep a sanitary pad in her backpack and to rehearse with her what she will do if she gets her period for the first time when she is at school or otherwise away from home.

It might help your child if you share some of your own experiences about going through puberty. Most of us didn't sail through this period of dramatic changes—it's unlikely that your daughter will. If a mom or aunt can share with a girl how she felt about her body changing, this may be reassuring. You might share information like:

- When I was your age, my body looked like_____.
- My biggest concern about my body during puberty was_____.
- Would you like to hear the story of my first period?_____.
- I remember boys were_____.
- I remember thinking sex was_____.

Books can be very helpful at this time. Your daughter may have a lot of questions as her body develops and changes during the next few years. It helps if she has some books of her own so that she has a reference she can look up in private to have some of her questions answered. Some of the books I like are listed in the appendix.

You might also want to look at some of these books together if that feels comfortable to you. My sister read *What's Happening to My Body: A Book for Girls* aloud with her ten-year-old daughter Emily. In our home, Alyssa preferred to have these books as her very own personal guide. It would not have worked to try to read them to her. Be sensitive to your daughter's own style. There are several books listed in the appendix that can help you.

There are also some excellent videos on puberty aimed at preadolescents. Again, you can give these to your daughter and suggest that you watch them together—or that she can watch them alone. If your child seems reluctant to watch them, you might want to tell her that they are available near the VCR and that she can watch them alone or with her friends. The appendix also lists some videos that I like.

You also might want to call around and see if a local Scout

troop, church, school, or community agency offers "preparing for puberty" programs. A good program will involve both parents and daughters, and will address not only the biological aspects of puberty but the emotional and social changes as well.

Single fathers with daughters who don't have mothers involved in their lives often feel especially awkward preparing their daughters for puberty. As one father asked me, "How can I talk to her about getting her period when I don't know what it's like?" Depending upon your relationship with your daughter, there are several ways to approach this. You could provide the education directly, just like a mom would. You could ask her aunt or grandmother or a close friend to take on this responsibility. You can definitely make sure she has books and videos to turn to with her questions.

I think it's important for parents to think through how they will respond to their daughter's first period. My first period started out as a pretty scary experience: I had gotten my period in school for the first time and panicked. I went to the school nurse, who gave me a pad and then sent me home from school early (which seems inexplicable to me now). When I came home from school and told my mother I had gotten my period for the first time, she slapped me lightly across the face! I was stunned. It turns out that this was an Eastern European custom (of unknown origin), but it felt like I had done something wrong. I burst into tears. Luckily, I also have a more pleasant memory of that day: My mother finally agreed to let me go buy Yardley Slicker (a type of very "in" lipstick in 1968, for you younger readers) because, she said, "I guess if you're old enough to have babies, you're old enough to wear lipstick."

I believe that young people have too few rites of passage these days, and I encourage you to consider treating first menstruation as a time of celebration. Some feminists have even developed rituals for celebrating first periods. In our home, we celebrated Alyssa's coming-of-age with a pair of pearl earrings that I had bought months before for this purpose and a "welcome to womanhood" dinner with our family. Think about how you might want to mark this passage.

I also want to talk about a sad thing that happens to many girls around puberty: The adult males in their lives—fathers, uncles, grandfathers—stop hugging, kissing, and holding them. All of a sudden, these men see the girl's changing body and withdraw physical affection. Some dads even stop hugging their pubescent daughters altogether. And their daughters sometimes feel ashamed, embarrassed, confused, and hurt by this rejection.

They didn't ask for this change in the relationship. It happened overnight. And a subtle message is given: All touch with a grown woman is sexual. Many adult women sadly remember a definite change in their relationship with their own father at this time. Remember, your daughter is still your little girl, and she still needs you. Work at trying to stay connected and affectionate.

Tanner Stages for Girls		
Tanner Stage	**Pubic Hair**	**Breasts**
1	None	None
2	Sparse	Small breast buds
3	Darker, begins to curl	Breasts and nipples enlarged
4	Coarse, curly, less than adult	Continued breast development
5	Adult triangle	Mature, nipple projects

Preparing Your Son for Puberty

Your son, like his sister, is also likely to notice that his body is changing before you do. He sees the stray hairs under his arms; he sees the hairs forming around the base of the penis and the scrotum changing color. He may wonder if he is normal. He may be concerned that it looks like he is developing breasts.

Many parents aren't privy to these changes. In fact, many parents are surprised when one day they spot their son's pubic or underarm hair or hear his voice crack.

Puberty in boys begins on average during the sixth or seventh grade, but some boys begin at nine and others don't begin until almost fourteen. The first physical sign is usually an increase in the size of the testicles; unlike breast budding in girls, parents are unlikely to observe this. During puberty, the penis will double in size. By Tanner Stage 3 (see box on page 158), ejaculation has usually occurred and there may be some production of sperm. In fact, some sexologists call this "spermenarche," to correspond to the first period (menarche). Most often, this happens to a boy through masturbation or as a nocturnal emission, or wet dream. By Tanner Stage 4, males are fertile and able to cause a pregnancy. The average length of time for a boy to go through puberty is three years, but it can vary from two to five years.

And if girls aren't being prepared adequately for menstruation, boys are definitely not being prepared for their first nocturnal emissions. I only have one male friend who was prepared by his parents for his first wet dream. Others had heard of it from older friends, often in a derogatory sense. A few grown men have told me that they remember thinking they were dying, had cancer, or more commonly, that they had wet the bed. Many remember being in a complete panic about what to do with their pajamas and bedsheets. For many boys, first ejaculation will take place during masturbation; see page 174 for more on these types of discussions.

You need to prepare your son for puberty, and as with your daughter, the Big Talk probably won't work very well. It might even be more difficult. So, try to be observant of the changes your son is going through and give information little by little. Talk about the physical changes he is going to experience and about his feelings. Be sure to cover the changes in penis and testicle size and the possibility that as he develops, he may have wet dreams. Tell him that you will understand if you come across stained sheets and that he can also just put them in the hamper. Or better yet, have him start doing some of his own laundry, a good skill for later in life.

You might want to ask your pediatrician for an assessment of your son's development at his annual physical. Your pediatrician can tell you about the stage of your son's development. He or she might help address some of your son's concerns.

Penis size can be a very big issue with preadolescent boys. They are comparing themselves to other boys at the urinals and in the locker rooms. In a recent class I taught of eighth graders, questions about penis size came up almost weekly in our anonymous question box. Boys this age need to know that the average adult penis is two to four inches flaccid and five to seven inches erect. They need to know that some men are "showers" and some men are "growers": Men who are smaller in a flaccid state have penises that grow longer during sexual excitement; bigger flaccid penises grow less. And they need to know that for most adults, penis size doesn't matter in sexual satisfaction.

Boys also need some information about erections. If you've been doing this since preschool—"sometimes penises are hard; sometimes they are soft"—they probably don't need much more than "Erections are more frequent as you go through puberty."

But if you haven't been talking all these years, this might be a more lengthy discussion. As boys go through puberty, they become more interested in sex and more likely to experience sexual attraction. And their penises are likely to become erect more often. In fact, most eighth-grade boys probably have several erections a day. These erections are often quite unexpected and unwelcome: The person in front of him bends down to pick up a fork in the lunchroom, a pretty girl in a short skirt gives a presentation right before him in English, a sexy advertisement appears in a magazine, and, vroom! There it is again! Boys need to know that this is a perfectly normal response, and that the erection will go away by itself. They also find it useful to hear about strategies for not letting any one else know that they are aroused: moving the lunch tray or books to a strategic place or holding the paper you are reading at waist level all help.

It might help your son if you or a man close to him shared some of your own experiences about going through puberty. You might share information like:

- When I was your age, I thought my body was_____.
- My biggest concern about my body then was_____.
- My first wet dream was_____.
- I remember thinking girls were_____.
- I remember thinking sex was_____.

Significant numbers of boys are also troubled by a condition called "gynecomastia." During puberty, they begin to develop an increase in the glandular tissue around their breasts; in fact, these boys sometimes are very worried and fearful that they are growing breasts or turning into a girl. Nearly one in five ten-year-olds will have this kind of tissue development.

In rare cases, this may be the result of diseases that are rare in young people: liver disease, hypogonadism, hyperthyroidism, or hypothyroidism. It could also be a sign of drug use. But before you panic, know that most of the time this is just a variation of male pubertal development, and usually resolves in a year to a year and half. It will continue past two years in fewer than one in ten boys. Still, your son's pediatrician will be helpful in ruling out serious underlying causes and reassuring your son that this type of breast tissue development is normal and that it will get better. (If your son is one of the rare boys who experience an extreme case of breast development, surgery is an option. But it is reserved only for extreme cases. Talk to your son's doctor.) It may be a good idea to tell your daughters about this as well: I once overheard my daughter and a friend giggling about a boy who looked like he was getting breasts. It helped them be a bit more empathetic when I told them how common it was, but how uncomfortable it made the boys.

Books and videos can be very helpful. You can buy books for your son to read together and/or read alone. It helps to have books on his bookshelves that he can refer to privately when questions arise. And as with your daughters, just because they don't seem thrilled to have these books given to them, doesn't mean that they won't look at them. (Some parents have told me that they have bought books on puberty and just left them laying around, hoping that their child will find them. I don't recommend this. Remember, you want your child to think that you are an askable parent; give them the books as a present!)

You might want to call around and see if a local Scout troop or church offers "preparing for puberty" programs for boys. These may be less available than programs for girls, but some Scout troops, Boys Clubs, churches, and other community agencies do have them. A good program will include boys and parents, and will address the upcoming physical as well as social and emotional changes.

Single mothers with sons often feel especially awkward at

this time. After all, these women have told me, "I've never had a wet dream or worried about penis size." Of course, you can talk to your son about these changes, just as you would other important issues. Or maybe you can ask your son's uncle or grandfather or a family friend to help you. Books and videos can help, too.

Tanner Stages for Boys		
Tanner Stage	Pubic Hair	Penis/Scrotum
1	None	Childlike
2	Sparse	Scrotum turns red and grows larger; penis childlike
3	Darker, begins to curl	Penis length increases; scrotum continues to enlarge and darken
4	Coarse, curly, less than adult	Penis increases in length and circumference
5	Adult	Adult

For Boys and Girls

It is important that your child know that the other gender also experiences the changes of puberty, and that many of these changes are the same for boys and girls. But I also think that boys need to know at least something about menstruation and girls need to know something about erections. I remember being mortified that a boy might see the sanitary napkin I was carrying in the sixth grade. Talking to boys and girls about the other's physical development—at least in general terms—is important for setting the stage for treating both sexes with respect and adult relationships. Too many of my women friends report that their husbands are unable to go to the drug store to buy them tampons! And knowing what the other sex is experiencing may cut down on inappropriate teasing or comments, a subtle form of sexual harassment. (See the Special Issues section on "Hostile Hallways" in this chapter.)

A Few Words about Early and Late Developers

Although most young people go through puberty right before or right after the start of their teen years, some begin puberty much earlier and some go through it much later. Both early and late developers face special challenges.

There are some children who begin puberty much earlier than nine or ten. The world's youngest mother is believed to have been a six-year-old Peruvian Indian—but rarely, some children enter what is called "precocious puberty" in the first year of life. Some girls begin pubertal development as early as age three, and although it may not be an indicator of a medical problem, this should be discussed with a physician. Interestingly, this seems to be largely the result of genetic coding in girls, and most often there is no underlying disease. Conversely, precocious puberty in boys is much more likely to be an indication that something is wrong. If parents observe pubertal changes in a child younger than seven, they should see a health practitioner for a medical evaluation as well as for possible interventions.

Less rare, though, are the children who enter puberty at around the age of eight—third grade! In fact, a recent study found that this is much more common than many parents and practitioners had thought. As many as one-quarter of African

American girls and about 10 percent of white girls actually begin puberty by the age of seven.

Early-developing girls have special problems. Because they look more mature, adults often expect them to act more mature. A third- or fourth-grade girl with developing breasts may feel awkward and want to hide her body. She may be relentlessly teased by her male and female classmates. She may indeed feel that she isn't normal.

Many parents pretend not to notice their young daughter's changing body. This may leave your daughter confused and feeling ashamed. Talking with her will help. Be sure to assure your daughter that the changes her body is going through are normal. Explain to her that the start of puberty is like an alarm clock that is preset to go off. Everyone begins at different times; her clock just went off early. Reassure her that there are probably other girls in her class going through these changes as well.

Try to relate to her feelings of embarrassment. Share stories of your youth when you felt that you were different or ahead of the crowd. Find an older woman who was also an early developer to talk with her. A little empathy can go a long way to making her feel better about herself. It can also help to rehearse with her what to do if she is the victim of sexually harassing comments at school or in your neighborhood. And you should also know that early puberty is one of the predictors of early first sexual intercourse and make sure that you are monitoring her social interactions.

Early-developing boys seem to have fewer problems. Although they may be teased for being taller, they are more likely to be socially accepted. In fact, many research studies point out that early-maturing boys are more likely to be popular, good in sports, and leaders in their schools. Still, other studies indicate that they may feel more insecure about their bodies and not ready to take on the more mature role that adults expect of early developers. If your son seems to be taller than his male classmates, and since you are unlikely to observe how his genitals are changing, it is a good idea to start talking about puberty earlier than you might have planned. You should also be sure to talk with him about how he is feeling about his early development. He still needs reassurance, even if he isn't bringing it up.

Boys actually have many more problems when they are late developers, compared to girls who seem to be more troubled by early development. Late-developing boys may be relentlessly teased, and may seem more childish and awkward. Many men I know who were late developers still remember their feelings of embarrassment.

Most of the time, early developers and late developers do not need any medical attention. But sometimes this is the sign of a medical problems that can be dealt with. In boys, if there are no signs of pubertal development by sixteen, medical intervention should be sought. In girls, if there are no signs by fourteen and a half, medical intervention might be appropriate. But some people go through puberty as late as nineteen or twenty. This is known by doctors as "constitutional delay of growth and adolescence." Seek a physician's advice.

Also talk to your child about their feelings. Late developers—both girls and boys—may feel awkward, shy, or embarrassed. They may wonder if their bodies are *ever* going to develop. They may not understand their friends' growing interest in boys or girls, and wonder what's the big deal? On the other hand, they may become virtually obsessed with their body and every tiny indication of development. It helps to have them talk to an adult who was also a late developer, and to encourage their own talents and skills. Point out that the world's best gymnasts are teens who are late developers. Assure them their bodies will change, just on their own individual time clock.

Important Messages about Puberty for Boys and Girls
- Puberty begins and ends at different ages for different people.
- Everyone's body changes at its own pace.
- Most changes in puberty are similar for boys and girls.
- Girls often begin pubertal changes before boys.
- Preadolescents only feel uncomfortable, clumsy, and/or self-conscious because of the rapid changes in their bodies.
- The sexual and reproductive systems mature during puberty.
- People are only able to have babies after they have reached puberty.
- During puberty, girls begin to ovulate and menstruate, and boys begin to produce sperm and ejaculate.
- During puberty, emotional changes occur.
- During puberty, many young people begin to develop romantic and sexual feelings.

"Am I Normal?"

Early, late, and average developers *all* share a common concern: *Am I normal?* Those rapid body changes can be very confusing. Nine- to twelve-year-olds can become obsessed with their appearance. In fact, psychologists say preteens have an "imaginary audience": They think that everyone is looking at them. Getting dressed and preparing their hair and faces can take hours. (Also see the section on body image and dieting on page 189.) They need your reassurance that they are attractive, and they need to know that it is not true that everyone is looking at them.

They often wonder, "Am I normal?" They are concerned about their height, breast size, or penis size, and whether they are too developed or too underdeveloped. They wonder whether their feelings, which sometimes fluctuate immensely, are normal. It is important to reassure your child often that they are developing according to their own personal, genetically predetermined clock. When they ask you questions like, "How come I don't have my period yet?" or say, "I hate being the shortest boy in the class," know that the question behind the question is "Am I normal?" You can say something like:

> *Parent*: I'm guessing that you are wondering if you are normal. But there isn't anything wrong with you. All boys and girls develop at different ages and different rates. By the time you are eighteen, you will be at your adult size. Until then, know that your body is developing at the right rate for you.

And you might want to share your stories, or even a photo, of you at this age. I went through a particularly awkward, chubby stage right before I got my period for the first time. Alyssa enjoyed looking at photos of me at that age. She said to me, "Boy, did you go through a bad puberty!"

Emotional Development

As any parent of a preadolescent will tell you, physical development does not occur separately from emotional or social development. The physical changes in our children are often much easier to accept than the emotional and social ones. Parents of nine- to twelve-year-olds often wonder, What happened to my nice, sweet, obedient child? Where is the child who couldn't wait to do things with me? How come they never seem to want to do anything with us anymore? How come they don't want to be seen with me?

Dr. Bob Corwin, a New York pediatrician, has said, "During childhood you are the hammer; during adolescence, you are the anvil." What he meant was that until adolescence, it is our job to teach, teach, teach, and during adolescence, sometimes we just need to be there. In the words of my friend and colleague Dr. Bob Selverstone, "It is fine for the child to walk away from the parent—you are preparing them for a life on their own. You must assure them that you won't walk away from them."

The reality is that during preadolescence and the years following it, conflicts with parents peak. Yet, it is important to know that only about one in six teenagers and parents experience a severe disruption in their parent-child relationship. Preadolescents are just beginning to separate from their parents, but they are also still looking to their parents for guidance and support.

These changes in the parent-child relationship are hard for everyone—your child and you. As Alyssa passed through this stage, I was struck by how much more difficult it is to parent your own adolescent than it is to work with other people's children. The teens in my congregation generally think I'm very cool; my daughter would often think I was embarrassing. At a recent annual meeting of the Society for Adolescent Medicine, the professional association of adolescent health-care providers, a luncheon titled "Surviving Your Own Child's Adolescence" was packed—and we work with adolescents every day of our professional lives! One day, Alyssa summed this up well. She turned to me after an argument and said, "Mom, you only think you're an expert in my age group!"

There is great variability in adolescent psychosocial growth and development. Just as children begin and end the physical stages of puberty at different times, so psychosocial changes occur differently in different children. In fact, what can be very confusing is that physical, social, emotional, and intellectual growth can all occur on very different timetables. Or all on the same day. Dr. Selverstone writes, "This move away from parents and toward friends is neither smooth nor easy. Parents are often confused: one day their teens insist on making their own decisions and the next day they beg for help. These shifts can confuse both the parent and the child. But it is a reality in the lives of preteens and teens."

Early adolescence has generally been characterized as occurring between the ages of twelve and fourteen, but many young people will begin these changes between ages nine and twelve. Try to remember that the goal of adolescence is adulthood: emotional, psychological, and financial independence from parents. And so, as these struggles begin, remind yourself that you really do want your child to grow up to be an independent adult. After all, do you really want your twenty-eight-year-old still living in your home, dependent on you for everything? Sometimes when things get tense with my daughter, I mutter to myself, "This is good. I want her to grow up to be an independent adult. She has to be doing this!"

The rapid physical changes of puberty described in the beginning of this chapter initiate many of these psychosocial changes. Adolescents are beginning a move toward greater independence and often are very self-conscious of their bodies. Fluctuating hormones often lead to extreme moodiness and unexplainable feelings. One minute, they want you to help them desperately; another moment, they are yelling, "You just don't understand" and slamming the door to their rooms. They may be less interested in doing things with you, maybe even being seen with you. (Many parents of sixth and seventh graders bemoan this: "Last year, she couldn't wait to do things with me, now I have to drop her off a half a block from the school.") Children of this age become very sensitive to what they perceive as adult criticism or advice, although they can be relentless in their criticism of parents ("Mom, you're not really going to wear that to the school concert?" is one line I heard recently).

They need to test your authority, and sometimes the authority of teachers and other adult figures in their life. This may include testing out values that are different than yours. I remember deciding at the age of twelve that I was a nihilist. I wore only black clothing, wrote dark poetry, and developed an inexplicable passion for Herman Hesse and Ayn Rand. Fortunately, my parents knew it was likely to be a passing phase and chose not to comment, and it passed pretty quickly. The research literature demonstrates that most teenagers go through a period of seeming to reject their parents' values, yet almost all of them adopt values very similar to their parents' by the time they reach adulthood. Try not to be alarmed when your child announces that they are a Buddhist, a socialist, a capitalist, a Republican—whatever is contrary to your family's expressed social values. It's likely to pass.

Friends become very important during the preadolescent years. Being popular is of utmost importance. The good news is that needing to be just like one's friends, known by psychologists as "conformity with peers," peaks in early adolescence and than starts to decline. The bad news for parents is that peer pressure can be intense during this period, and usually dictates the type of sneakers that are acceptable, the length of one's skirt and pants, and even whether or not it is still cool to be seen with one's parents. It also can lead to experimentation with cigarettes, alcohol, and yes, sex.

Cliques often emerge during this period. Depending on where you live, there are the jocks, the nerds, the skaters, the preppies, the hippies, the populars, the burnouts, and the freaks. Ask your children what the groups in their school are called. One look at a preteens' clothing is often enough to place them in a peer group. And just like when we were in junior high school, there is probably the "in" or "popular" group that sometimes lords over all the others. Your child may for the first time feel pressure to engage in certain activities to be part of a group, and for some young people, that means experimenting with smoking, shoplifting, or cheating on tests. Among thirteen-year-olds, a surprising 22 percent have smoked a cigarette; 10 percent have used marijuana; and one-third say they have tried alcohol.

Your preadolescent child is likely to pick friends who are similar to him or her in social status, intellectual ability, and interests outside of school. And then the influence of their friends makes them become even more similar. Adolescents are more likely to spend more time with their friends and classmates than either their parents or their parents' friends. And study after study shows that preadolescents are happiest when they are with their friends. This can be hard on parents, who long for the earlier closeness of early and middle childhood. Remember again, you want them to become independent of you by the time the teen years are over. Relax and try to enjoy your child's newfound delight in their friends.

It's a good idea to talk to your preteen—a lot—about friendships and peer pressure. It can be especially important but difficult if you think that your child is getting in with the wrong crowd. Try to keep the lines of communication open without being dictatorial. Certainly if you feel a crowd is harmful, you will want to limit your child's contact or try to divert them with other social groups at your church or religious organization or with sports teams. Be sure your child knows that a good friend makes you feel good about yourself—that someone who asks you to do something contrary to your values is not a good friend. You can ask them about which qualities they value in a friend, and what about them makes them a good friend.

You also want to make sure that your child has a peer group. Some preteens experience tremendous loneliness and alienation. They don't seem to have the skills to make friends at school, or for some inexplicable reason, they don't seem to fit in. It's important that they have friends their own age. If your child doesn't seem to find friends at school, help them seek alternative peer groups. Encourage them to join after-school activities, sports teams, musical groups, Scouts, church or temple youth groups—whatever interests them. And talk to them about how to start and continue friendships.

One of the biggest challenges for parents of children in these preteen years is what to do when your child seems to be falling in with the wrong crowd or has a friend you don't like. This may be your last chance to intervene in these kind of situations. Certainly, by the time your child is driving, there are not many ways to control their friendships. But you can exercise control now: Just like when your child was younger, you can decide whose house you will let your child go over to, who she can go to the mall with, what he can do after school, who she can talk to on the telephone and for how long. But unlike when your child was a toddler, it is better to discuss these issues and seek your child's cooperation than it is to "lay down the law." If you are uncomfortable with a particular child, invite them over for dinner and watch their interaction with you and your child. Trust your instincts; don't be afraid to tell your child that you prefer them not to spend time outside of school with this child and tell them *why* you feel this way as well. But try to help them reach this conclusion without making it a showdown.

Just as when your children were younger, try to get to know your children's friends' parents. This can be more challenging than when you were all hanging around the playground together. But it is one way to know more about your child's friendships, and it also lets your preteenager know that you are staying involved in their life. These parents can also be additional caring adults in your child's life.

During these preteen years as your child asks for increased privileges, you may also hear, "But Jen's mom lets her!" After months of hearing this about everything from lipstick to going to the movies on her own, I invited Alyssa's five best friends' parents over for a meeting. The goal was to see if we could agree on a set of rules relating to clothes, makeup, and going out for all of our daughters. After all, "Jen's mother lets her!" doesn't work when you and Jen's mother have talked about these things together. And although our families' religious values were very different—we ranged from Unitarian to non-churchgoing to Pentecostal—in an hour, we were able to agree on a set of principles for our seventh graders.

Think about trying this with your preteen's friends' parents. All you need is a few hours and a list of issues you are dealing with in your home. This might include telephone privileges (we decided that calls would be limited to ten minutes, and that there would be no phone calls after 8:30 P.M.), makeup (we decided lipstick was okay, eyeliner was not), and dating (we said no). It could also include which videos are acceptable for slumber parties (G only? PG? PG-13?), going to the mall, going downtown, going to movies alone, and going to boy-girl parties.

Your child at this age is also beginning to develop a sense

of identity. The ability to think abstractly has just begun. Children at this age frequently daydream and turn inward. Daydreaming is normal and actually quite healthy. Trying on roles in one's head is frequent. Giving your child a diary or journal and respecting their right to keep it private can be very important to encourage your new adolescent to think about and explore feelings. Many parents worry that their preadolescent child is spending hours alone in their room; try to remember that this time alone can be an important part of growing up.

Children at this age can seem very dramatic. They often feel that they are continuously onstage. They may become convinced that their problems are unique and that no one understands them. "Mom, you just don't understand" followed by a slammed door is not uncommon. Be sure to be there to talk when the door is opened again.

Early Adolescence at a Glance
- Puberty is the defining event
- Adjusting to pubertal changes
- Concern with body image
- Beginning to separate from parents
- Increased parent-child conflict
- Presence of social group cliques
- Identification with peer groups
- Concentration on relationships with peers
- Concrete thinking, but beginning of new ability to think abstractly

The Preteenager with No Questions

Some preteens make puberty education easy. They are full of questions, eager for discussion with parents, and delighted with the changes in their bodies. With others, it's much more difficult.

In the Passover seder, the ritual retelling of the exodus from Israel, there is a passage in which four questions are asked. One of the questions is, "And what about the child who never asks?" And the Haggadah answers, "Then you must tell them." And it is the same with sexuality.

There are some preteens who don't ask questions about sexuality. There are some who are embarrassed to even have the issue raised. One of my friends, for example, reports that his twelve-year-old daughter is embarrassed just by the words "underpants" and "bra." And if you haven't been talking to your child about sexuality before this age, you very well may have given them a clear message that you are uncomfortable addressing these subjects. Unfortunately, it's going to be hard at this point to make them think differently. But even if they don't ask, you should assume that they have questions.

Do not, for a minute, think that the child who doesn't ask knows all that they need to know. They may just be shy about these issues. They may think that they should know everything by this point. They may not know how to ask you about this subject. You may have inadvertently told them that you are not comfortable talking about sexuality issues.

This is the age at which some parents think they should give the Big Talk. Resist this urge. It's unlikely to do much more than reinforce how uncomfortable you both are talking about sexuality issues. Remember, it is better to have many small conversations than to bombard them with facts. In fact, facts are less important than giving your child the message that you are willing to talk with them about these important subjects.

And don't just leave books lying around the house and hope that they will read them. An unbelievable number of parents have told me that this is how they handle their preadolescents' need for information. As Bob Selverstone wrote, "Leaving books around does not mean children will (a) read them, (b) understand them, or (c) remember later and not need to talk more." It is better than nothing, but it further reinforces the message that you really don't want to talk to them about sexuality.

So, what do you do? Remember that your child probably does have questions and concerns, and they may not know how to bring them up with you. It is your job to take the lead.

The easiest way to begin, as at all the other ages described in this book, is to look for the teachable moment. Use a situation on television to discuss your values about sexuality: "Do you think Mary should be dating that boy who has a child? I don't think she should have been in the bedroom alone with him." And wait for your child to answer. Sometimes it is easier for preteens to talk about fictional characters than it is to talk about themselves or their friends.

Listen to a teen-oriented radio station when you are in the car together. I can almost guarantee that a song will come on with a sexual message. When it is over, ask your child to tell you the lyrics. Discuss her reactions to them. Share your reactions. Share with her some of the lyrics to songs that played when you were her age. (Remember the Rolling Stones' "I Can't Get No Satisfaction?") Or wait for a news story with a sexually related theme and talk about your reactions and values.

Driving your child and their friends in the car can be an interesting insight into teen issues. I am sometimes surprised when my daughter's friends are willing to talk about issues in the back seat as if I am deaf sitting up in the front! It can be a good way to catch up on what your child is really thinking and identify issues for future discussions. I also find that in our home bedtime is an easier time to talk about important issues than after school.

One suggestion some people have found helpful is to ask your preteen to explain something to a younger brother, sister, or cousin. "I'm wondering Sue, if you can help me talk to your little brother about Aunt Carolyn's being pregnant" or "I want to talk to Dave about AIDS tonight; can you help me out?" In this way, you are letting your child know that you know they have some information about these types of issues, and you are giving them a face-saving way to listen to your descriptions. It is a great way to correct any misinformation they have and to let them know that these are topics you talk about in your home.

Another idea is to talk with another adult about one of these issues while your child is present. Letting them listen in on a discussion on sexual harassment lets them know that you think they are grown up enough to handle these topics. Inviting their opinion lets them know that you value and respect their ideas. And it lets them know that adults still have to work through many of these issues.

But what if you've never discussed these issues before? Well, despite my strong belief that you need to start in infancy, I also believe it is never too late to start. Resolve today that you will begin. At dinner tonight, take a deep breath. "Honey, I've been reading a book about talking to children about sexuality. I know we haven't talked much before. I'm sorry. But it's important that we do. And I'm hoping we can begin, little by little." And then wait. You may be surprised by their reactions.

Masturbation

There are some subjects that are difficult to address, no matter how open you have been previously with your child. Masturbation may be one of these.

During puberty, many boys and girls begin to masturbate for sexual pleasure. Research shows that as many as three-quarters of boys and about half of girls under the age of fifteen masturbate. And unlike the more casual genital touching I've talked about in previous chapters, many of these young people are seeking orgasm and pleasure. Masturbation is often the first way that young people—and sometimes adults—experience orgasm or ejaculation.

Boys, in particular, seem to take up masturbation at puberty. Remember *Portnoy's Complaint?* Parents of adolescent boys often call me concerned about the amount of time their son seems to be spending locked in the bathroom or in their bedroom. The boys I have taught in this age group are often worried that they are masturbating too often and wonder whether their desire to masturbate is normal. And some girls have these same type of questions too.

My colleague Sol Gordon taught me that the answer to "How much is too much?" is "Once is enough, if you don't like it." And if masturbation is interfering with your child's school, homework, or friends, or your family life, it is too often. Otherwise, frequency is highly individual. Some adults masturbate every day; some once a year. And some people never masturbate. In fact, your preteenager needs to know that although many preteens, teens, and adults masturbate, some never do. And if they are worried about this behavior, they should talk to you or another adult they trust.

Some families oppose masturbation, and if you feel this way, it's important that you communicate your feelings and values to your preadolescent. However, I do not know of any research data to show that forbidding masturbation will stop young people from engaging in this behavior, although it might increase their guilt. Regardless of your beliefs, every preadolescent does deserve to know one fact: There is no evidence that masturbation causes physical or mental harm.

Going Together

Of course, preteens are not only interested in solitary sexual behaviors. Many of them are beginning to explore their first relationships, and some of them are beginning to experiment sexually with another person.

Today's teenagers don't date. They don't go steady. They "go together." And they go together as early as the fifth grade. Middle school seems to be the beginning of "going together." If your middle school starts in fifth grade, going together starts in fifth grade. If it starts in seventh grade, that's when this starts.

What does it mean? Well, it's probably not very different than when you were in junior high school. It means that you told your friend that you liked someone, they told that someone, who either said yes, they liked you too or no, they didn't. And if the answer was "yes," then you were going together.

And, most of the time it probably does not mean much more than that. "Going together" during the preteenage years is a dress rehearsal for future teenage romantic relationships. "Going together" does not mean going out on dates. One of my friends reported a funny conversation with her twelve-year-old daughter in which the mom kept asking, "going together *where?*" There is no "where" for these couples. For most of them, going together means having everyone at school know that you are boyfriend and girlfriend, and dancing together at the school dances. It is unlikely to mean dating, going to each other's houses, or experimenting with sexual behaviors beyond kissing. It is likely to mean little beyond long telephone calls and exchanging notes in the hall. And it's likely to be very short-lived. One-quarter of these relationships break up in less than three months; almost all do in a year and a half.

Talk to your child about what the prevailing norms are in their school. Are some of their classmates going together? Have they ever gone with someone? Wanted to go with someone? Nationally, about half of young people ages twelve to fourteen have been in a romantic relationship.

You might be surprised to find out that many fifth and sixth graders think it is better to not tell their parents that they are going with someone. Some girls have told me, "My parents would forbid it." Some boys have told me, "My dad would just make fun of me." And other preteens have told me that this is a private issue and none of their parents' business.

But I think that this is an important area for discussion. It gives you an opportunity to be part of your preteen's life and to talk with them about their feelings. Many preteens worry that they are not going with anyone. And sometimes there is a great deal of peer pressure to have a boyfriend or girlfriend.

One friend of mine's daughter is quite worried that she's in the seventh grade and still has not had a boyfriend. Another friend's son is worried because he seems to change girlfriends every week. And indeed, these relationships are often quite short-lived.

Parents often worry about these first relationships because they are concerned that they will lead to sexual experimentation. Some parents indeed do forbid them. I generally believe that most of these relationships are harmless and a way for young people to experiment safely with romantic feelings. Since they generally do not involve out-of-school activities, forbidding them is unlikely to be successful. All it is likely to do is teach your child to go behind your back—not a good foundation for successfully parenting an adolescent.

Although many of these going-together relationships are more about the status of having a boyfriend or girlfriend than anything else, for some young people, they are about first love. Kids in one of my classes speak wistfully about their friends Bonnie and Max; they are the only couple in their group who are really in love. When I asked one what this means, one of my students said, "They just are, Debra." And then she told me that Bonnie and Max spend time together every day after school, talk on the phone every night, and have even been known to sneak the portable phones into their bedrooms so that they can listen to each other when they sleep. (By the time I finished revising this chapter, Bonnie and Max had broken up.)

Messages for Preteens about Love
- A person can show love for another person in many ways.
- Liking yourself enhances loving relationships.
- People are capable of giving and receiving love.
- Love is not the same as sexual involvement or attraction.
- The feelings of "falling in love" are different from those in a continuing relationship.
- In a mature love relationship, people encourage each other to develop as individuals.
- First love is often one of life's most intense experiences.
- Love is a difficult concept to define.
- Knowing for sure if you are in love can be difficult.

In surveys and in classrooms, one of the questions that preteens and teens most want help on is "How do I know if I'm in love?" First love can be both exhilarating and frightening. It can also be all-consuming. Your child may be spending hours daydreaming or talking on the phone with their boyfriend or girlfriend.

Before you read any further, I want you to shut your eyes and think for a minute about your first love. What was his or her name? How did you feel about him or her? How did you feel when they called you? Smiled at you? Held your hand? And how did it feel to have adults not take you seriously or call this most important special relationship "puppy love"? Was there someone who did not return that first or second infatuation? How did that feel?

Most of us can remember those feelings quite intensely. And for many people that first love may be the most intoxicating, consuming love of their life. And for sure, it is one we never forget. I'm fond of telling the audiences I speak to that although I have been happily married for almost two decades, I have never written Ralph's name one thousand times on a napkin the way I did my first love, who incidentally contacted me when he read a review of the first edition of this book.

If your child is in love, it is important to take their feelings seriously—and it is just as important to share your values and set limits. This is an opportune time to reinforce that feelings are different than behaviors. Ask them if they want to hear about your first love. Be sure to share with your child which behaviors you think are acceptable for a twelve-year-old and which behaviors are not. Help them understand that being in love is only one part of a relationship; it also includes trust, intimacy, and friendship.

Help them differentiate between what my colleague and friend Sol Gordon calls Immature Love and Mature Love. He advises young people that mature love is "when your caring about the other person is just a little more important to you than having the other person care for you. The relationship is mutually enhancing and energizing. Immature Love is when the other person's caring for you is a lot more important than your caring for the other person. Your love is a burden on the other person, and the state of being in love is exhausting." In other words, when you are in a mature love, you feel great and you are nice to the world. When it's immature love, you feel obsessive, depressed, and cranky to everyone around you. Teenagers (and indeed many adults) experience both kinds.

Use your child's crush or first love as an opportunity to talk about your values about sexual behaviors and young people. Explain that love and sex are different; they often they go together, but not always.

Most preteenagers do little more than kiss. But those first kisses can be thrilling. Remember spin the bottle? It is still a staple at those first boy-girl parties. But romantic kisses are usually reserved for boyfriends and girlfriends. If your child is going with someone, they are most likely exchanging kisses.

But a minority of preteens move beyond kissing. Here are some facts that may surprise you:

- Seventy-three percent of girls and 66 percent of boys thirteen and younger have kissed.
- Twenty percent of thirteen-year-old boys have touched a girl's breasts, and 25 percent of girls this age have had their breasts touched.
- Twenty-three percent of boys and 18 percent of girls this age have fondled someone's genitals.
- Seven percent have had sexual intercourse, a number that has been decreasing steadily since 1991.

There have been many media stories about oral sex in middle schools, often alarming parents to no end. I have received calls from parents and the press about alleged instances of oral sex at bar mitzvah parties and on school buses. The fact is that there is almost no research on this question, so the stories are just that—stories. It is no surprise that national research studies have not asked middle-school children, "Have you received or performed oral sex?" It's hard to imagine any middle-school principal allowing a researcher to ask that question. In a recent study of inner-city teens, only 3 percent of the virgins ages twelve to fourteen said that they had had oral sex. In another study of young people who are involved in a faith community, 11 percent of the boys and 14 percent of the girls in the ninth and tenth grades had had oral sex but not penile-vaginal intercourse. And based on discussions and writings by teenagers themselves, oral sex in middle school almost always involves girls performing it for boys, not vice versa.

My sense is that the media reports are overstated and sensationalized. I frankly can still remember the name of the girl that was offering oral sex in my eighth-grade class. My sense is that oral sex is probably still "gross" for almost all upper elementary school-age young people, and engaged in by less than 20 percent of middle-school teens. And that the profile of the young teen that gets involved in these behaviors is probably consistent with the young teen that begins having intercourse: They feel disconnected from parents, school, and friends, and learn to use sex as a way to feel powerful and connected.

The fact is that parents can make the difference in whether their child is involved in mature sexual behaviors. Clear, direct messages from parents about the importance of abstinence are vital for young people in this age group. In one national study of young people in grades seven to twelve, perceived parental disapproval of adolescent sex was strongly related to teenagers' remaining abstinent. An even stronger correlation of delay of first sex was the teens' sense of a strong parent and family connection—believing that their parents loved and cared for them. In fact, teens who are close to their parents are more likely to postpone involvement in intercourse, have fewer sexual partners, and use contraception consistently if they do have intercourse than young people who are not as close.

It is part of your job as a parent of a preteenager, and then as a parent of a teenager, to set limits for your child's dating. Too many parents, perhaps in an attempt to encourage their child's move to independence, stop being involved. You need to decide when, if, and under what circumstances your child can date. Some professionals believe that it is inappropriate for young people to begin steady, one-on-one dating much before age sixteen. Certainly we know that early dating is more likely to lead to earlier sexual experimentation than if dating starts later. Others believe that this type of dating rehearsal is harmless, but that parents should still set limits for where the child is allowed to go: "You can go to the movies in the afternoon, but I will drive you and pick you up."

You need to decide what is right for your family. But it may help you to know that the key seems to be parents' supervising and monitoring behavior, not the actual limits that are set upon the child. Teens whose parents closely supervise them are more likely to be older when they first have intercourse, to have fewer partners, and to use contraception if they do have sexual intercourse. Being too strict, however, can backfire. Some studies suggest that the "forbidding" approach doesn't work: Teens with "very strict" parents are *more* likely to become pregnant. The key is staying involved and communicating your expectations to your child.

And be sure to talk with them when they do come home from their group and single dates. Ask them open-ended questions like "What was the most fun part of your date?" not "How was it?" (I can just hear the twelve-year-old mumbling an answer to this yes/no question: "Fine," "Okay," "Boring.") Try not to judge them. The better you are at listening, the more likely your child is to want to talk to you.

Although the limits are up to your family's values, I do want to share with you one bias I have that is supported by a great deal of scientific literature. Middle-school children should not be allowed to "go with" or "date" someone who is more than two years older than they are. The research is very clear: Young teens who date older teens (and sometimes, horrifyingly, adult men, or, rarer, adult women) are much more likely to get involved in high-risk behaviors: drinking, doing drugs, having sex, and even getting pregnant. Just think of the news story of the pregnant female schoolteacher and her eighth-grade student. Know who your child is going out with, and let them know, before it happens, that they are not to go out with people more than a couple of years older than they are.

Parents also need to talk about abstinence with their children. Your preteen children need to know what you think about teens' having sexual intercourse, and the behaviors you expect of them. They need to understand that abstinence is the only 100 percent effective way to prevent pregnancies and sexually transmitted diseases (STDs). They need to know which sexual behaviors you think are appropriate for someone their age: For most of us, that means stopping at kissing and hand-holding.

No matter what your opinion is about premarital sexual relationships, I cannot think of a single parent or professional who thinks that middle school–age youth are ready for a mature sexual relationship that includes intercourse. And the facts bear this out: The earlier a teen begins intercourse, the less likely she or he is to use birth control, and hence, the more likely she is to get pregnant. S/he is also more likely to have multiple partners in adolescence. And if she is a girl, it is more likely that her partner is significantly older: Among young mothers aged eleven and twelve, the father is, on average, ten years older.

Tell your child your values about intercourse in clear terms:

Parents: Young people your age are too young to have intercourse. Our family believes that intercourse should wait for (pick one or several) marriage/college/a mature love relationship/adulthood. I hope we will talk a lot about this as you start to have romantic relationships.

But, this abstinence message does not mean that you don't have to talk to your children about birth control and sexually transmitted diseases. In today's world, it is vital that by middle school, young people have some basic information about contraception and STD prevention. They need to know about teenage pregnancy, sexually transmitted diseases, and AIDS.

Now, you may be thinking, doesn't our school teach about this? It may, but in most parts of the country, information about birth control and STD prevention often does not get addressed until high school, if at all. And although your child may seem knowledgeable about these issues, in my experience, preteens' knowledge is often quite superficial. Yes, they have grown up hearing about contraception and AIDS, but they may know little about how people actually protect themselves against pregnancy and diseases.

I am often asked if talking to preteens and teens about contraception and condoms at the same time you are promoting abstinence doesn't give them a "double message." My response is that we give young people double messages all the time. Think about it for a minute: For years, you have been telling your child to use sunscreen and play outside when it is nice out.

Is that a double message? And in a few years, you will teach your children to drive cautiously and wear their seat belts. Is that a double message? And not to drink, but if they do drink, not to drive. Is that a double message?

Telling your children that you hope they will abstain in high school, in college, or before marriage, depending upon your family values, and then being sure that they know about contraception and condoms is the same. I unequivocally do not want my son or daughter to become prematurely involved in sexual intercourse; I gave Alyssa very clear messages that this is adult behavior and it shouldn't be part of high school dating relationships. But I've also told her that I love her and care about her health and her future, and if she decided on something different, I wanted to be sure that she would protect herself and her partner.

Your preadolescent needs to have some basic information about contraception and condoms. They need to know that there are many different methods of contraception, but only latex and polyurethane condoms provide protection against sexually transmitted diseases. They need to know that some contraception requires a visit to a health-care professional and a prescription (birth control pills, Depo Provera, IUDs, diaphragms, and cervical caps), and others can be obtained at the drugstore, supermarket, or convenience store without a prescription (condoms, foams, gels, and suppositories). They need to know that every method has advantages and disadvantages, and that no method is 100 percent effective.

They also need to have some very basic information about sexually transmitted diseases, and as I pointed out in the last chapter, you can assume that they definitely have heard about AIDS. They need to know that there are diseases that are only transmitted through unprotected sexual contact, including sexual intercourse and oral sex. They need to know that abstinence is the very best protection against STDs, and that condoms are the only contraceptive method that provides such protection. They also need to know that you don't get STDs from toilet seats, but you can get herpes from kissing someone with cold sores or through oral-genital contact.

Messages for Preteens about Sexually Transmitted Diseases and AIDS

- There are many types of sexually transmitted diseases.
- Sexually transmitted diseases include gonorrhea, syphilis, HIV, chlamydia, genital warts (HPV), and herpes.
- To have AIDS means that HIV has done enough damage to the body that certain serious diseases have been acquired.
- Oral sex can transmit sexually transmitted diseases, including HIV.

But most importantly, I think, preteens need to know that you want them to come to you if they are even beginning to think about the possibility of having sexual intercourse. Let them know you want to help them think through this decision. Tell them your values about premarital sex, but also be sure they know you are there for them regardless of what they decide. And if they can't or won't come to you, they should talk with another trusted adult, like a family friend, aunt or uncle, or minister or rabbi. Stress that this is an adult decision and that they need an adult's help in thinking it through if they are really ready. (I go into much more about adolescent sexuality in *Beyond the Big Talk,* the sequel to this book.)

> **Messages for Preteens about Abstinence**
> - Young teenagers are not mature enough for a sexual rela-
>   tionship that includes intercourse.
> - Abstinence from sexual intercourse is the best method to
>   prevent pregnancy and sexually transmitted diseases,
>   including HIV.
> - Young people who go with someone need to discuss sexu-
>   al limits with their boyfriend or girlfriend.
> - Abstinence has many benefits for preteens and teens.

"Mom, I Think I'm Gay"

Not all preteenagers are interested in young people of the
other gender. Some young people simply haven't developed yet, and so have no interest in dating or relationships.
Others at this age begin to struggle with their feelings of sexual attraction to people of the same sex.

According to a study by University of Minnesota psychologist Gary Remafedi, more than one in four 12-year-olds say
that they are unsure about their sexual orientation. In today's
world, with its more frequent images of people who are gay and
lesbian, many young people have questions about homosexuality and bisexuality. And a significant number of young people
wonder about it for themselves.

Sexual orientation usually emerges during the teenage
years. By the time they reach the age of eighteen, 95 percent of
teens are sure about their sexual orientation. As with heterosexual activity, studies show that most gays and lesbians have
their first same-sex sexual experiences by the time they are
twenty. And a majority of adult gay men report that they felt
"different" as children, although most did not "come out," or
share their orientation with others, until early adulthood.

Yet there is a difference between being gay or lesbian as an adult and being unsure of your orientation as a young teenager. It is common for preadolescents and adolescents to be attracted to people of the same sex, to develop crushes on a teacher or coach of the same sex, and even to experiment with sexual behaviors with a friend of the same sex. That doesn't necessarily mean that they are homosexual. But some young people are gay or lesbian.

So, how do you respond to your child's proclamation, "Mom/Dad, I think I'm gay." First, take a deep breath. You are likely to be having a rush of feelings: You may be scared, confused, worried, or upset. Even the most liberal heterosexual parents I know have told me that their initial response to their child's coming out to them was confusion. One woman confided in me that she was surprised by her reaction: "We have gay friends, we march in gay pride parades, so why am I feeling so disappointed?" How much more difficult it must be in homes where parents have strong religious or political feelings against homosexuality.

I defined sexual orientation in the chapter on preschool-age children. Remember, sexual orientation is about more than sexual behavior. It is defined by whether you fall in love with, are attracted to, have fantasies about, or engage in behaviors with someone of the same sex or someone of the other gender. It is important for you to know that who we will feel sexually attracted to and fall in love with as adults is not a "choice" or a "preference." It is an orientation and as much a part of us as being tall or short or left-handed or right-handed. And no one really knows why a person has a particular sexual orientation. Many studies have been done on it. Some scientists believe it is genetic or based on prenatal hormones, others think it is a combination of physical factors and social factors, and still others think that it is a combination of all of these.

Find out what your child knows about homosexuality and why they may be feeling this way. You could say, "Tell me what makes you think that" or "Tell me what you know about homosexuality." Really try to listen to their answers. Resist the urge to say, "You're too young to say that" or worse, "We will never accept that." You can tell them that many young people have these feelings, but most people don't really know their sexual orientation until they are older.

But the most important thing your child needs to know is that you will love him or her no matter what. All children need their parents' unconditional love, support, and acceptance. If your early adolescent is discovering that she or he is homosexual, they need to know that more than ever.

You need to gauge what to do next. Does your child want to continue the conversation or do they want some time to think about it? Are they confused or scared or are they happily telling you something important about themselves? Are they asking you for additional help? Counseling may be helpful for your child and for you as you deal with these issues. According to the American Psychological Association, counseling cannot change someone's sexual orientation. It will not work to take your child to a counselor in order to have them change their orientation! But counseling can help you and your child deal with your feelings.

There is an excellent organization for parents of gay, lesbian, bisexual, and transgendered children of all ages. PFLAG stands for the Parents, Families, and Friends of Lesbians and Gays. Their phone number is (202) 467-8180. They have chapters in many places around the United States, as well as a Web site, www.pflag.org. It may help you to talk with other parents who are having this experience.

Body Image, Appearance, and Dieting

Almost all preteenagers, regardless of their sexual orientation, begin to be obsessed with their appearance at some point. Not only do they worry if their bodies are normal, but they also obsess about their clothes, hair, and skin. A single pimple can ruin an entire day. A bad haircut is reason enough to want to stay home from school. Having the right clothes reflect one's status in the desired peer group.

When Alyssa was in sixth grade, she wore only loose flannel T-shirts and baggy jeans, despite having other neater clothes in her closet. One day, when we were going to a school concert, I asked her if she couldn't wear something a little nicer. She summed up the prevailing norm in her school this way: "Mom, we have three styles in our school: dorky, schleppy, and slutty. Aren't you happy I'm schleppy?!" How could I argue?

Remember that styles of dressing are one of the ways that preteens announce to the world their social status and that they are growing up. Think back for a moment about how you dressed during your teen years: Are low-rise tight jeans and midriff tops any more objectionable than hot pants or peasant blouses worn without bras? Did you fight with your parents about the length of your hair or your miniskirt? Is an earring or purple hair on your son really worse than your teenage Afro or a ponytail? (And did you outgrow these teen fashions all by yourself or are you still wearing them to your job today? Those of us who are baby boomers certainly have little reason to worry about peek-a-boo midriffs, even if we don't like the way it looks!)

The point is that clothing—and music—are ways that preteens and teens express themselves and their culture. And frankly, we adults are not supposed to like it—or get it. Just like with our toddlers, we need to pick our battles; you don't want to be fighting with your child all the time. You need to decide what your standards are, and communicate which clothing and appearance behaviors you will and won't accept. In our home, I let both Alyssa and Greg know that they could not make any permanent changes to their bodies until the age of eighteen: only one ear piercing (but stick-on additional earrings and fake belly button rings are okay), no permanent tattooing (although temporary stick-ons or henna are okay), no permanent hair dyes (but hair mascara is okay). It might be a little difficult if one decided to shave off all of his or her hair, but I would remind myself that it was temporary, bite my tongue, and live with it—until the next style came along! In fact, I made it a point to compliment my daughter's bright blue nails, fake tattoos, and jewelry as often as I could sincerely.

My only other restriction has to do with overtly sexually provocative clothing and shoes. My colleague and friend Bob Selverstone wrote, "The line between what looks good and what is sexy is often hard for young people to make. Until their sense of self is more mature, young people may seek attention from dressing and acting 'sexy.'" I am sometimes appalled at middle-school events when some girls are wearing ultra-short skirts, midriff shirts with visible bras, and too much eye make-up with platform sandals: The message these young Lolitas are giving to adults may be unintentional but all too real.

Talk to your child about which messages they want people to take from their appearance. Ask them what messages different styles send to important adults like teachers and grandparents: "What do you think that boy with the pierced nose and chin is trying to say to adults?" Ask which messages they want to send with their appearance.

Try to make this a dialog, not a monologue. Yelling or even saying nicely, "Go change" or "Take that off right now" is unlikely to help your child learn to make good decisions or improve your relationship with them. Talking about these issues and striving to reach a compromise will. For example, you may hate the baggy, low-slung jeans look on your son, but it is probably not harmful. A compromise, for example, might be that he can have two pairs to wear to school, but that he wear corduroys or khakis for church and visits to grandparents.

The preteen and teen concerns about clothes is really a concern about fitting in and appearance. During these preteen years, it can be hard for young people to feel attractive. They are bombarded with media images of young, sexy, thin men and women, and they may wonder why their appearance seems so far away from these media images of perfection. Skin problems, awkwardness, and body fat in the wrong places are real manifestations of puberty.

Help your child feel as attractive as possible during puberty. When I look at seventh-grade pictures of pudgy me, I can't help but think how much difference a good haircut or more attractive pair of glasses might have made. Help your child feel as attractive as possible during this changing time. If they are having problems with their skin, and you can afford it, consider a trip to a dermatologist who can probably solve all but the worst cases of adolescent acne. Encourage your child to use good hygiene: Regular showers and hair washing can help appearance as well as prevent those emerging body odors.

To the extent that you are able, buy your child clothes that are considered acceptable by their classmates. Let them do chores around your home to earn the $125 for that pair of must-have tennis shoes. Or help them find jobs baby-sitting, mowing lawns, or doing errands for a neighbor. At $5 an hour, they will learn the value of money, you'll get chores done, and they may even decide the clothes are not worth the money!

But also do remind your preteen that looks aren't as

important in life as character, intelligence, love, and friendships. Make sure your preteen knows that although they can't change how quickly they go through puberty, they can work to develop their minds, manners, and friendship skills.

The extension of young people's appearance concerns is, unfortunately, dieting, binge eating, and more seriously, eating disorders. According to a 1997 study of preadolescent girls by the Commonwealth Fund, an unbelievable 39 percent of fifth- to eighth-grade girls report that they have been on a diet. Even more horrifying, 13 percent of girls this age reported that they had binged and purged. Of these, 36 percent said that they did it at least once a day, and 11 percent said they did it a few times a week. Boys can have eating disorders too; increasingly, young men are becoming concerned with being toned and "buff."

It is critical that you talk to your children about healthy nutrition and body image. Eating disorders are often the result of poor body image and low self-esteem, and sometimes control issues between parents and teen children. If you suspect that your preteen is severely limiting food or binging and purging, you and they need professional help. (See the appendix for organizations that can help you with referrals.) But almost all preteens need help with accepting and living with their bodies.

Think about the messages you are sending to your preteen about food. If you or your partner is constantly dieting, you are not sending them healthy messages about body image. Many young women, in particular, grow up in homes with mothers who give the message that a woman's worth is somehow tied up in how thin she is. When you put down your own appearance or body size, you model that it is okay not to like yourself. For some reason, white girls on average seem to be more concerned than African Americans and Latinas with what nutritionists have labeled the "drive for thinness." (Remember that bumper sticker: "You can never be too rich or too thin"—well, you can be too thin. It's called anorexia.) In some homes, men are also obsessed with working out, six pack abs, dieting, and marathon running. Let your child know that a person's appearance is determined by heredity, environment, and health habits. They can't affect how tall they will grow, or their inheriting Aunt Alice's thick calves, but they can eat wisely and exercise regularly. And remind them that puberty is a time of great physical change, and that they will probably feel happier with their bodies in a few years.

Acknowledging your child's feelings about their appearance and their body is important. Letting them know that there have been times when you weren't happy with your looks can be reassuring. And talking about what's really important about a person also can help. Know that preteen children, both boys and girls, who are teased or criticized about their weight or who have lower self-esteem are more likely to get in trouble with eating disorders. So please don't tease or criticize your child about their appearance and certainly try not to nag them about food or eating.

The opposite of the dieting, too-thin child is the overweight preadolescent. According to the U.S. Department of Health and Human Services, 14 percent of preteens are overweight. It is very difficult to be very heavy at this age. And how a parent responds to a chubby preteen may determine whether this child accepts him- or herself or their body, or begins a lifelong struggle with food and self-hate. It is important for body-conscious parents to know that weight gain is to be expected during puberty. The fact is that some bodies are just large, and helping your children learn the difference between body size and healthy eating is crucial. Your large child may never be thin, but they can eat well and exercise regularly. Fitness and size are not necessarily the same. If you are concerned that your child is unexpectedly heavy, please talk to a nutritionist or pediatrician before putting your child on a diet.

My colleague Maureen Kelly of Planned Parenthood of Ithaca, New York, says that you should never tell an overweight preteen (or adult for that matter), "But you have such a pretty face." It is important to never make a connection between your child's weight and their self-worth: Never say something like, "You'd feel so much better about yourself if you lost a little weight." Help your child feel attractive, and try to find them clothes that are similar to what their friends are wearing. In a 2003 article, Maureen wrote, "Never, ever, ever say anything that supports narrow concepts of body or health… expressions like 'But, I'm sure she's pretty on the inside' undercut everything you are trying to teach."

The National Eating Disorders Organization (see the appendix) has developed a list of things parents can do to help prevent eating disorders in children, including:

1. Avoid negative statements about your own body and eating.
2. Educate yourself and your children about the genetic basis of differences in body shapes and weights. Be sure they understand that weight gain is normal and necessary during puberty.
3. Take your children seriously about what they say, feel, and do, not how they look.
4. Scrutinize how your child's school portrays women in the curriculum and whether there are posters, books, or contests that reinforce the myth of thinness.
5. Make sure your child knows the difference between body

shape and personality or value. Don't allow them to use such phrases as "fat slob," "pig out," or "thunder thighs."

6. Teach children the dangers of dieting, the value of moderate exercise, and the importance of eating nutritious foods. Do not dichotomize foods into "good/safe" foods and "bad/dangerous" foods.

7. Encourage your child to be physically active and to enjoy what their bodies can do and feel like.

8. Do not use food as a punishment or reward. Make family mealtimes relaxed and friendly.

Messages for Preteens about Body Image

- A person's appearance is determined by heredity, environment, and health habits.
- The way a body looks is mainly determined by the genes inherited from parents and grandparents.
- The media portray "beautiful people" but most people do not fit these images.
- Standards of beauty change over time and differ among cultures.
- The value of a person is not determined by their appearance.
- Eating disorders are one result of poor body image.

Teen Magazines

Many people think that one of the causes of poor body image in preadolescents are the media images that they are surrounded by. Teen magazines are written for teenagers—right? Wrong. Although these magazines have names like

Seventeen, YM and *Teen*, and their presumed target audience is young women ages thirteen through seventeen, many middle-school girls read them avidly. In one study, almost two-thirds of middle school–age girls read these teen fashion magazines, and three-quarters of these girls said that they are an important source of fitness and beauty information. Teen boys turn to *Sports Illustrated* (especially that swimsuit issue!) and even *Men's Health* and GQ.

These magazines often contain articles that don't seem appropriate for children twelve and under. In my recent trip to the newsstand, articles included "Sexy Beauty Secrets," "The Ultimate Get-a-Guy Guide," and "10 Boy Truths You Need to Know." Many parents worry that the information in these magazines will put ideas into children's heads and that they will want to experiment with adult behaviors. Others, like me, worry about their emphasis on attracting a boyfriend and about the endless advertisements of anorexic-looking young models in skimpy clothes.

Let me tell you about a call I received from a good friend of mine with a twelve-year-old daughter. "Kimberly is reading YM magazine and this month there was a long article about contraception. It was completely unnecessary for her and I'm afraid it's going to give her ideas." I asked her, "What kind of ideas?" She answered, "She'll want to have sex so she can use a contraceptive method." And then she stopped, realizing how unlikely that sounded. After all, she then admitted, "People don't have sex to try out a condom!"

But her concern underlies a question that many adults have: Does talking to children about sexuality or exposing them to information cause them to experiment with sex? The answer is a resounding "No!" There is a large body of research going back more than twenty years that shows that having sexuality education courses does not cause young people to have sex. And although I don't know of any studies that looked at the relationship of preteens' reading teen magazines and having intercourse, I'm pretty sure that there would be no correlation. At worst, your child is likely to be turned off by an article about sex, skip it, or think it's silly. At best, they will remember it when they are older and face some of these dating situations.

Think about yourself for a minute. What do you do when you are reading a magazine and it has an article that doesn't have anything to do with you? Most likely, you skip it, or you read it to learn something about others. I know that for years I just skipped the articles on perimenopause because they didn't apply to me—now I read all of them!

Other people, particularly women like me who gave up *Glamour* for *Ms.* magazine, are concerned about the sexist images these magazines give girls. And indeed, they are rife with articles on how to catch a cool guy, how to wear the right makeup and the right clothes, and ads that promote a very narrow standard of beauty. However I was pleasantly surprised as I reviewed magazines for this update. In addition to the beauty and popularity tips, this month's magazines had articles on nutrition, handling stress, gang prevention, and puberty.

There are many new 'zines on the market for girls with a feminist flavor as an alternative to *YM*, *Seventeen*, and *Twist*. They include *Teen Voices Online* (www.teenvoices.com); *Blue Jean Online* (www.bluejeanonline.com); and *New Moon: The Magazine for Girls and Their Dreams* (800-381-4743, www.newmoon.org).

The trick is getting your preteen daughter to want to read these alternative magazines. To be honest, I had no luck in this area. Alyssa's day was made when her aunt Pat gave her a *Teen* magazine. Her comment about the alternative 'zines were,

"Mom, you read them!" Perhaps the trick is to start your daughter's subscription to them at nine, and hope she becomes a loyal reader. I know some ten-year-olds who love *New Moon*. Or ask a friend to order your daughter a subscription and hope she reads it when it comes in.

Parents of preadolescent boys may have a different issue about magazines: Their sons seem to have skipped straight from *Boy's Life* to hidden copies of *Playboy* and *Penthouse*—or online sex sites.

"Dad, Will You Buy Me a *Playboy*?"

Not only do preadolescents become obsessed with their own bodies, but boys, in particular, become extremely curious about females' bodies. In a class I once taught, eighth graders were asked to fill out an anonymous questionnaire about their concerns about sexuality. Every boy checked "to see a naked woman." Only one of the girls checked "to see a naked man."

This may surprise you considering the media images that flood our air waves—MTV, VH1, X-rated sites on the Internet, and the almost weekly Victoria's Secret catalog that shows up in the mail. It seems much easier these days to see half-clad bodies than when many parents were growing up; many men my age report scouring *National Geographic* for pictures of women's breasts or the *New York Times Magazine* for women in their underwear. And I have learned from men and women who are a few generations older than I am that they used to wait avidly for the new Sears Roebuck catalog for its section on women's bras and panties and men's briefs. This curiosity is a natural part of growing up and becoming a sexual person.

A friend of mine recently called me to talk about his ten-year-old son Spencer. Spencer shared with my friend that the boy next door had told him that his dad kept a stack of *Playboys* in the attic. He asked my friend where he kept his *Playboys*. My friend answered that he did not have any. His son's next question shocked him, "Well, Dad, can you go buy me one?" My friend stammered out that he would think about it and get back to him.

Then he called me. "Debra, I didn't know what to do. Help!"

I asked him to tell me more about his son. It turns out that Spencer had always been pretty interested in sexual issues. When he was six, he had asked his father, "Do you and Mommy f***?" Although only ten, he is already developing pubic hair and an interest in girls.

I suggested that my friend think about what might be behind Spencer's question. Did Spencer have any books that included pictures or diagrams of adult nude women? (No.) Did he think Spencer had looked at the *Playboys* next door? (Not yet.) Did he worry that Spencer might have been involved in some type of sexual activity? (No.)

His answers made me think that Spencer was simply expressing age-appropriate curiosity. Yes, he had seen Victoria's Secret catalogs but he had never actually seen a woman nude. He did not have any of the puberty books that I recommend in the last chapter. He also probably was feeling some competition from the savvy boy next door. I then asked my friend what values he wanted to convey to his son about erotic materials. I suggested he and his wife discuss some issues together:

- What do we think about *Playboy*?
- What messages are we comfortable giving our preadolescent children about erotica?
- How do we feel about the types of women who are portrayed in *Playboy*?

- Are we comfortable having sexually explicit materials in our home?
- Would we prefer other children to introduce our child to sexually explicit materials?
- Are we comfortable being the ones to provide *Playboy/Playgirl* to our son/daughter?

For many preadolescent boys, this interest in sexually explicit magazines reflects both their curiosity and a desire to do something "grown up." In fact, some parents face this situation in a less direct way. It is not uncommon for a parents of a sixth or seventh grader to one day, while cleaning their son's room, find a *Playboy* or *Hustler* magazine hidden under the mattress.

This is a classic "Oh no, what do I do now," situation. So let's go through the steps. This is a good time to review your own and your partner's values about sexually explicit material. One in five adults say that they used sexually explicit materials in the past year. In fact, it may surprise you to know that more than 600 million X-rated videos are rented from neighborhood video stores each year. Nevertheless, four in ten adults say that these kind of materials should be illegal.

I believe that adults should have the right to make up their own minds about sexually explicit materials, and that they should be available for personal use. I strongly oppose, however, the use of violence, degradation, and exploitation in such materials and believe that the portrayal of children is always reprehensible.

I also believe that parents of children have a responsibility to protect their underage children from this type of material. Your nightstand is not the place to keep your erotica, no matter how convenient that might be for you. The bookshelves are not a good place to keep your X-rated movies unless you want the neighborhood children seeing them. And I do not believe that early adolescents are ready for these images.

So if you and your partner enjoy these materials, talk together about where it is safe to keep them from your children. Maybe you have a safe in your home and only the adults know the combination. Maybe there is a place in the back of the closet under the out-of-season clothes. Maybe you want to throw them all out for the next few years.

But, you will also want to think about your values about your preteen child's having access to these materials. Are you comfortable with his/her first views of adult nude bodies being the air-brushed, perfect bodies of models? Do the pictures reflect the values you want to pass on to your child about adult male/female relationships? Are there other materials in the magazines—columns, advisors, comics—that are not age appropriate?

Communicate these values to your child. You can set limits on the kinds of materials that are acceptable in your home:

> *Parent:* I find these magazines sexist and degrading to women. In real life, women and men do not have these types of perfect bodies. I'd be happy to share with you some books that I think will answer your questions.

Then think about how to respond to your growing child's curiosity about naked bodies. Some of the books in the appendix for preadolescents have explicit drawings. Know that if you forbid these materials, it doesn't mean your child won't see them. They'll just turn to their friends, an older friend, or even the Internet. The important thing is to keep the communication lines open.

The Internet

Playboy is tame compared to the types of materials your child can probably find on the Internet. More than thirty million children under eighteen use the Internet. And although I firmly believe that the Internet offers children wonderful opportunities, it also poses new sexuality challenges for parents: How do I keep my child safe on the Internet? and how do I keep my child from viewing sexually explicit images?

In some homes, the children may know more about computers and the Internet than the adults do. So the first thing you have to do, is learn more. If you are not using the Internet in your job, you may want to take a course at your local library or adult education program. Spend time online: become familiar with what's available on your online system, what type of parental controls there are, and whether there are kids-only forums. Going on the Internet with your child is a way to learn more about its advantages and drawbacks, and is a way to introduce the Internet safety concerns I will discuss. There is a terrific Web site—www.netsmartz.org—that can help you and your child.

Surfing the Web is a wonderful way for your child to obtain up-to-date information on an unimaginable number of topics. It is helpful for schoolwork, developing important research skills, playing games, and listening to and seeing cultural treasures. It is also the easiest way for a person of any age to access sexually explicit materials. And it may provide a child molester with access to your children.

At any search engine, your child can type in "sex." I was actually shocked the first time I did this. At Google.com there were 260 million matches for the word "sex," one thousand times more than when I first did this in 1998. On the first page, I found "free sex photos" and "free porn thumbnail galleries."

On this cover page, you see such phrases as "oral sex," "anal sex," and "porn sex." (And trust me, I have left the most offensive words out.) Clicking once on one of these sites led me to photos of nude couples having intercourse and oral sex.

Now to be fair, if I had "parental controls" on my system this site would have been blocked out. Parental controls have been developed to help you. AOL, Yahoo, and Google all have controls for parents as part of their services. These controls allow parents to restrict access to chat rooms, email, and the Internet when children are using the computer. Ask your service what controls are available. Although it is possible that these controls may block out some information you would like to have your child have access to, I think it is wise to consider them for your home. It is hard for me to believe that an eleven-year-old boy surfing the Net for "sex" would be deterred by the screen telling him that he was not allowed to enter!

Commercial products are also available. CyberPatrol, Net Nanny, and CyberSitter are some of the parental controls that are probably available at your local computer store. You should carefully review these controls to see if they meet your family values before purchasing.

It can be confusing to pick the right filter. There is (of course) an Internet site that can help: www.GetNetWise.org. It has a questionnaire you can use to identify which features you want to control: There are filters that limit the amount of time your child can spend online, deny access to hate groups and violence, and block the ability for children to give out identifying information. According to GetNetWise, "Almost all filtering software blocks sexually explicit material, including depictions of sexual acts, nudity, and sexually explicit text." There are special browsers for children that do not display inappropriate words or images and tools that limit search engines to those that are "kid oriented," performing only limit-ed searches or screening search results. Interestingly, no tool does it all. When I put in all of these criteria, no filter matched all of them. When I put in what I personally would be looking for in our family, only two filtering tools emerged.

My concern about parental controls is that they also may block sites that might have information that you think is appropriate for your child to access. The values of the filtering technology might not be your family's values. I'm pretty sure most of them aren't mine. GetNetWise advises parents to check out "if the company publishes its criteria for filtering, its list of filtered sites, or the key words or phrases that it uses to block material." A study by the Kaiser Family Foundation found that when filters are set at their most restrictive levels, not only is pornography blocked but so are one-quarter of the health sites that deal with sexuality issues. Even when set at their least restrictive levels, filters block out about one in ten health sites from searches on the terms "condoms," "safe sex," and "gay." A filter may deny your child access to information, for example, from the American Cancer Society about breast and testicular self-exams. One of my minister colleagues report-ed that she was blocked on her home computer from searching for the word "compassion" because of the "ass" in the middle.

There are other, nonelectronic ways that you can protect your child from these materials. One way, similar to my advice above on televisions, is not to put a computer with online capacity in your child's private bedroom. There is simply no way that you can monitor its use if your child has unlimited access behind a closed door. Keeping the computer in the liv-ing room, family room, or study means that all use is observ-able. It is equally important to establish some ground rules.

Email and chat rooms pose other kinds of problems. Your

child can receive "spam" at their email address: unsolicited mail that can be sexual in nature, advertisements for sexually explicit sites, or commercial advertisements. In addition, your child will be meeting total strangers in chat rooms; they may be exposed to conversations that are not age appropriate, and unfortunately, there are some pedophiles who use the Internet to meet children.

Recently, a headmaster at one of New York City's most respected private schools was arrested on charges of soliciting sex from teenagers online. In October 2003, in one week in the Chicago suburbs, the police charged an elementary-school principal with distributing pornography and a pediatrician with possessing more than three thousand sexually explicit messages of children. In Westchester County (New York) alone, there have been 78 arrests of doctors, lawyers, priests, administrators, and coaches on child pornography and child solicitation. I have to admit it makes me queasy to even think about it.

There is no question that children are being exposed to inappropriate messages. In a national 2007 study, one in seven young people ages ten to seventeen had received a sexual solicitation on the Internet in the last year, and one in three had unintentional exposure to pictures of naked people or people having sex while online. The majority said that they had not told their parent that this happened.

There are a variety of things you can do to help ensure that your child is using email, social networking sites, and chat rooms safely. Some parents choose to share their email account and password with their children: The advantage of this is that you can read and screen all their email first; the disadvantage is that they can access yours! Some filters route your child's email to your account first; the question is whether you want to interfere with your child's private messages this way. If you don't read their snail mail, should you be reading their email?

Talking to your child about email may be their best protection. Tell them that you only want them to open up mail from addresses they know. Tell them to never open up an attachment. Ask them to show you any email that they receive from unknown sources or that make them uncomfortable. Set a limit that email is only to be sent and received from people they know. Make sure you know your child's password and tell them that you will periodically be screening their email messages to make sure they are following these rules.

Ask your children which Web sites they would like to visit and check them out together. Tell your child never to fill out personal profiles; indeed, Web sites for children are not allowed to ask for personal information without a parent's permission. I do not allow Greg to fill out any personal information forms; the fact is that they provide information that someone could use to identify and reach out to your child. Limit chat-room access to child-friendly chat rooms, or tell your under-twelve-year-old child that they are off limits completely. Children under fourteen should not be allowed on social networking sites like MySpace and Facebook. Greg asked for an instant message (IM) account at eleven; he was only allowed to IM his sister at college or his dad or me at work. If he wants to add someone to his buddy list, he has to come to us to talk about it first.

Help your child select their screen names. Nastygrl is not an appropriate screen name for a twelve-year-old, but neither is iluvtball. But, they are not as dangerous as a screen name like Eric12Boston, which gives out too much personal information. It is best to pick out a nondescript name that might be used by an adult as well as a child. Again, sexual predators are looking for children; if it's a chance that the person behind the screen name is an adult, they will go elsewhere.

Make sure your child knows that if they receive emails from

someone they do not know to just delete them. And if they tell you that they receive offensive or dangerous emails, you should save them, turn off the monitor, and contact your local police. You can also help investigations into possible online predators and child pornographers by contacting the CyberTipline of the National Center for Missing and Exploited Children (800-843-5678).

You need to tell your child, and repeat often, that they are never to meet someone offline without your permission and without you accompanying them. My inclination here is to just say no to any such meetings. There was a heartbreaking incident in Danbury, Connecticut, where a man in his twenties murdered a 13-year-old-girl he had met online and began to visit at neighboring malls. Your child needs to know that grown-ups can pretend to be children online and that such meetings are just too dangerous to consider.

The NetSmartz Web site has Internet Safety Pledges you can print for you and your child to sign about Internet use. Here are some of the rules that they suggest for grades three to six. You can copy them on a piece of paper and both you and child can sign them and post them next to the computer. Print-ready copies are available at www.netsmartz.org:

- I will talk with my parents or guardian so that we can set up rules for going online. The rules will include the time of day that I may be online, the length of time I may be online, and appropriate areas for me to visit while online.
- I will tell my parents, my guardian, or the trusted adult in charge if I come across any information that makes me feel scared, uncomfortable, or confused. I will not download anything without permission....
- I will never share personal information such as my

address, my telephone number, my parents' or guardian's work address/telephone number, or the name and location of my school...

- I will never respond to any messages that are mean or in any way make me feel uncomfortable. If I do get a message like that, I will tell my parents, my guardian, or the trusted adult in charge right away so that he or she can contact the online service. And I will not send those kinds of messages.
- I will never meet in person with anyone I have first "met" online without checking with my parents or guardian. If my parents or guardian agrees to the meeting, it will be in a public place and my parents or guardian must come along.

NetSmartz also has interactive online activities for your children to teach them Internet safety at different ages.

Finally, and on a nonsexual topic, you need to talk with your child about legal issues and the Internet. Martha Stansell-Gamm, chief of the Computer Crimes and Intellectual Property section at the Department of Justice, writes for *Newsweek* magazine, "Many kids who would never steal mail or CDs or destroy property think nothing of helping themselves to copyrighted music over peer-to-peer networks, or launching a destructive Internet virus." Your children need to know that they cannot download music, movies, games, or software without your permission and without you checking out that it is legal for them to do so.

Remember, though, that no control or rule is as important as your being involved with your child and their use of the Internet. My husband, Ralph Tartaglione, who has researched the Internet and children, wrote an article for parents in the

"SIECUS Report," and summed it up this way:

> It is, however, the parents themselves who must take the time to understand the technology and the role they must play in monitoring its use. To do this, parents must communicate directly with their children to acknowledge the risks that exist in this technology, to explain how to reduce these risks, and to set up ground rules, based on the family's personal values and belief systems.

Special Issues

Hostile Hallways

Unfortunately, sexual harassment is a fact of life in most middle schools. Sexual harassment in school is defined as "unwanted and unwelcome sexual behavior that interferes with the student's life." Sexual harassment includes unwanted sexual comments, jokes, and gestures; leaving another student sexual pictures, photographs, and notes; sexual graffiti about a specific student in the bathroom or locker room; spreading sexual rumors; flashing or mooning; touching, grabbing, or pinching in a sexual way; brushing against another student in a sexual way; pulling clothes off; blocking another student in a sexual way; and forcing kissing and other sexual behaviors. In a study conducted by the American Association of University Women in 1993, 85 percent of girls and 76 percent of boys in grades eight through eleven reported that they had been victims of one of these behaviors. One-third had been harassed by the sixth grade. And almost six in ten of these students report that they have done these offensive behaviors themselves!

Most sexual harassment in schools occurs openly in classrooms and hallways, rather than in secluded areas. Students are usually harassed by other students, but in one in five cases, they are harassed by adult employees. And students report that this type of sexual harassment affects their lives: Girls in particular report that they feel less confident and more afraid to go to school following incidents of sexual harassment.

Unfortunately, this same study found that students are not likely to tell adults about these incidents. Only 7 percent told a teacher and only one-quarter told a parent. It is important for you to talk with your child about whether any of these incidents has ever occurred to them and how they might handle such incidents if they need to. You can use a news story about sexual harassment to begin the discussion:

Parent: I read today that sexual harassment in schools is increasing. I'm wondering if anyone has ever bothered you in this way.

Schools are supposed to protect their students against sexual harassment. Schools need to have sexual harassment policies, with clear penalties for perpetrators, and these need to be communicated to students and parents. A clear procedure for handling complaints of sexual harassment should be established. Ask your middle-school principal or guidance counselor what the policy is and how it is communicated to students and teachers. Suggest forming a committee to draft a policy if one doesn't exist. These policies should be sent home to parents at the beginning of each new school year.

Your child needs to know that if a situation gets out of hand, you will get involved. Young people can learn to say "Get lost," but sometimes they need your help. When Alyssa was in the seventh grade, a fellow student began to harass her. It started with notes stuck in her locker, but quickly accelerated to unwanted phone calls, unwanted pizzas being delivered to our home, and threatening emails. Following the pizza and email incidents, I went immediately to meet with the assistant principal before school started and clearly stated that if the school couldn't stop this behavior, I would press charges against the young man involved. We were prepared to go to the police but wanted to first allow the school to handle it. Fortunately, a meeting with the boy, his father, and the assistant principal brought these behaviors to an end. But I have to admit that it was scary for all of us.

How to Evaluate Sexuality Education Outside of the Home

Most parents want help in educating their children about sexuality. In fact, the most recent public opinion polls show that more than eight in ten parents want the schools to offer sexuality education programs. Unfortunately, despite this widespread support, only 5 percent of young people receive comprehensive sexuality education from kindergarten through twelfth grade.

Now you may be thinking, "Slow down there. What's this stuff about kindergarten?" Well, I believe that sexuality education is like other school topics. The real goal is the development of behaviors for adulthood—whether that's being able to read a newspaper or a novel (reading), balancing your checkbook (math), voting (civics), or forming healthy adult intimate relationships (sexuality education). But just as children need to learn their letters and numbers in kindergarten in order to read English literature and do calculus in high school, they are helped by a foundation of basic concepts in elementary school to prepare them for their future decision making about sexuality.

Teachers in school districts that offer programs at each

grade level report that these children are better prepared when they are introduced to more complex topics like contraception and STD prevention in later years. One teacher has told me that you can actually sense the difference when a child moves into their district: "My sixth graders are ready to talk about puberty. You don't sense the embarrassment and you don't get the giggles and red faces. And you can always tell the children who just moved in and haven't had the earlier preparation."

But how do you know if the program is any good? A good sexuality education program is not just about sex. It is not only about anatomy and reproduction (what I call "plumbing lessons" or what one of my colleagues dubbed "an organ recital"), and it is not only about how to prevent the twin disasters of teenage pregnancy and sexually transmitted diseases. It encompasses the totality of sexuality: human growth and development, personal skills, relationships, sexual health, sexual behavior, and sexuality in our culture.

In 1991, SIECUS convened the National Guidelines Task Force to develop a model framework for sexuality education. Updated in 2006, they identified four key objectives for a sexuality education. Ask your school if they

- provide a comprehensive base of information. The chart on page 216 lists the 37 topics that should be included in a comprehensive sexuality education program.
- give young people an opportunity to examine their own values and attitudes, and hear the values and attitudes of others in an atmosphere that promotes tolerance for differences in beliefs.
- include opportunities to build personal and interpersonal skills, such as communication skills, decision-making

skills, assertiveness, and how to say no to peer pressure.
- help young people develop skills to exercise responsibility regarding sexual relationships, at the middle-school level and above, addressing abstinence, resisting pressures to become prematurely involved in sexual relationships, and using contraception and other sexual health measures if sexual activities do occur. Is there a good balance between abstinence and contraception and STD prevention information?

Find out who is teaching the program. In too many cases, the person who is assigned to teach the sexuality education class has had little formal training or insufficient background in this subject.

When my children take sexuality education, I will want to ask the following questions:

- Why was this teacher selected to teach this unit?
- Are they teaching it voluntarily?
- What academic training do they have to teach sexuality education?
- Did they major in health education in teacher's college?
- Are they certified to teach about sexuality?
- What recent inservice classes have they attended?
- Are they ever observed or supervised in the classroom by experienced sexuality educators?

I think it is important for parents to have the option of reviewing the curriculum and any materials that will be used prior to the beginning of the class. A parent's orientation night is usually a good idea, and you should make a point of attending and asking your questions. In fact, if your school has a program without one, I would ask them to hold a parent's session before the class begins. Here are some questions you might want to ask the teacher:

- Tell me about your background in teaching sexuality education.
- How will you handle children's questions about sensitive topics?
- Are there topics that are off limits?
- Will there be tests and homework? What will they include?
- What audiovisual materials will be used? Will I have the chance to preview them?
- Will you use a variety of teaching methods or will you just lecture and show videos?

Good sexuality education programs do not just lecture to young people. Professionals recommend that sexuality education address the three learning domains: cognitive, or information sharing; affective, or an opportunity to discuss and share one's values and feelings; and behavioral, or the development of skills. Ask whether programs will extend beyond lectures and videos. A good program will include such things as role-plays, group exercises, skill-building opportunities, group discussion, parent-child homework assignments, and times when the genders may be separated.

A good sexuality education program will seek to involve parents at home. Ask if there are to be homework assignments for parents and young people to do together. Ask if there will be a parent/child evening. And a good program will allow a parent who, after carefully reviewing the program, decides it is not what they want their child to receive, to opt their children out of the class but require the parent to cover the material at home.

Ask about the values and principles that provide the foundation for the program. "We don't teach values in our school" should not be a reassuring answer. Every sexuality education program should be based on a set of values, and I believe that these should be shared explicitly with the parents. Some of the values that I think are important to share, adapted from the list designed by the National Guidelines Task Force, are as follows:

- Sexuality is a natural and healthy part of living.
- All persons are sexual.
- Every person has dignity and self-worth.
- Parents are the primary sexuality educators of their children.
- Families should share their values about sexuality with their children.
- People should respect and accept the diversity of values and beliefs about sexuality that exist in a community.
- Sexual relationships should never be coercive or exploitative.
- All children should be loved and cared for.

And for children in middle school:

- All sexual decisions have effects or consequences.
- All persons have the right and the obligation to make responsible sexual choices.
- Young people develop their values about sexuality as part of becoming adults.
- Premature involvement in sexual behaviors poses risks.
- Abstaining from sexual intercourse is the most effective method of preventing pregnancy and sexually transmitted diseases, including HIV.

Topics to Be Covered in a Sexuality Education Program

- Human Development
- Anatomy
- Reproduction
- Puberty
- Body Image
- Sexual Orientation

Relationships
- Families
- Friendship
- Love
- Dating
- Marriage/Lifetime Commitments
- Raising Children

Personal Skills
- Values
- Decision Making
- Communication
- Assertiveness
- Negotiation
- Finding Help

Sexual Behavior
- Sexuality through Life
- Masturbation
- Shared Sexual Behavior
- Abstinence
- Human Sexual Response
- Fantasy*
- Sexual Dysfunction*

Sexual Health
- Contraception
- Abortion
- Sexually Transmitted Diseases, Including HIV
- Sexual Abuse
- Reproductive Health

Society and Culture
- Sexuality and Society
- Gender Roles
- Sexuality and the Law*
- Sexuality and Religion
- Diversity
- Sexuality and the Arts*
- Sexuality and the Media

*Source: SIECUS, *Developing Guidelines for Comprehensive Sexuality Education*, 1996.

Afterword

Being a parent today is tough. So is being a child. The reality is that today's children are exposed to sexual issues much earlier than ever before. And the AIDS epidemic has made sexuality education a life-and-death issue. It was difficult to go through puberty when we were in the sixth or seventh grade, and there are even more pressures on young people today. How is your child doing?

In 1994, SIECUS, the Sexuality Information and Education Council of the United States, convened the National Commission on Adolescent Sexual Health, a consortium of leading experts on adolescent growth and development. They developed a list of characteristics of a sexually healthy adolescent. I have modified and adapted this list to describe a sexually healthy child. Most of these behaviors apply from preschool through middle school; in fact, many of them apply to teenagers and adults as well!

Does your child:

- feel good about his/her body?
- know the correct names for *all* the parts of his/her body?
- if older than eight, know about pubertal changes?
- know that these changes are normal?
- understand that television and movies can give un-realistic ideas about love and romance?
- have good friends of both sexes?
- know what to do if a friend is pressuring him or her to engage in bad behavior?
- help out as a family member at home?
- respect the rights of others?
- show respect for teachers, youth leaders, and other adults?
- feel comfortable talking to you about sex-related issues?
- accept adult guidance easily?
- show empathy to others?
- know what to do if someone touches him or her inap-propriately or makes them feel uncomfortable?
- respect your right to privacy and his or her siblings' right to privacy?

Your child is unlikely to score a perfect 100 on this check-list—many adults would not score a perfect 100. But it is part of a parent's job to help them develop the skills to do so. You are helping them develop the foundation for a sexually healthy adulthood.

If your child seems seriously lacking in one of the above areas, address it with them. Consider seeking professional help if your child does not seem to have any friends, lies to you on a regular basis, or seems to engage in any behavior that doesn't seem age appropriate.

But also relax and remember to enjoy your child. Teaching

them about sexuality can be one of the joys of parenting. Yes, it can be challenging, but it can also be entertaining, funny, and heartwarming. It tells your children that you are there for them, even when the issues or situations are difficult.

Before Gregory was born, seven-year-old Alyssa asked me, "Mom, are you going to tell the new baby about all the stuff about babies and bodies you've told me?" I asked her what she thought. She answered, "I hope so. It's great we can talk about these things."

Remember, setting the foundation for talking about sexuality during your child's early years will make these talks much easier in adolescence when they become even more pressing. The exciting, challenging world of adolescence is just around the corner!

Appendix
Sources for
More Information

I hope this book has answered many of your questions and concerns. But I also expect that it raised some additional questions and some issues for you. I firmly believe in seeking out additional information. This appendix includes a list of some of my favorite places to go with questions and some books that you may wish to share with your children. It also includes a list of hotlines and Web site addresses.

Parts of the original list were adapted with permission from "Sexuality in the Home: A SIECUS Annotated Bibliography of Available Resources." I have updated it using the many excellent Web sites that are included.

WEB SITES

There are a variety of Web sites that have information for parents on talking to their children about sexuality. Here are some of the ones that I like the best.

www.advocatesforyouth.org
Advocates for Youth has developed a Parents Sex Education Center. In addition to lists of Web sites for parents and teens, the site includes articles by some of the country's best sexuality educators.

www.cfoc.org
This is a web site for the Campaign For Our Children in Maryland. It has a section for parents, including one on "how to talk to your school board about sex," and a section for teens.

www.drspock.com
This is the Dr. Spock Web site. I've written many articles about sexuality education for this site, and there are other articles by pediatricians that may be helpful.

www.familiesaretalking.org
The Sexuality Information and Education Council of the United States has developed this separate Web site for parents on sexuality education in the home.

www.plannedparenthood.org
Planned Parenthood Federation of America's Web site includes "Resources for Parents and Other Adults," as well as everything you want to know about contraception.

www.talkingwithkids.org
The Kaiser Family Foundation sponsors this site, which includes articles about talking with children and teens about alcohol, drugs, HIV/AIDS, violence and sexuality.

ADDITIONAL READINGS

Here are some of my favorite books that you can share with your children. All of these books should be available at your local chain bookstore or at Amazon.com. The books span the cultures' differing values about sexuality; be sure to review them before giving them to your child to make sure they support **your** *family values.*

SEX BOOKS

3- TO 8-YEAR-OLDS

What's the Big Secret? Talking About Sex With Girls and Boys
by Laurie Krasny Brown and Marc Brown (New York: Little Brown, 2000)
This was Gregory's favorite book about sexuality. It introduces the differences between boys and girls, and covers anatomy, reproduction, pregnancy, privacy, and birth.

How You Were Born
by Joanne Cole (New Jersey: HarperTrophy Publishers, 1994)
Colorful photographs explain birth in a simple, accessible way.

Did the Sun Shine Before You Were Born?
by Sol and Judith Gordon (New York: Prometheus Books, 1992)
This book is a classic. It begins at conception and ends with birth.

Happy Birth Day!
by Robie Harris (Cambridge, MA: Candlewick Press, 2002)
This book has lovely illustrations of a baby's first day, told by a parent in a reassuring tone, about a hospital birth.

Heather Has Two Mommies
by Leslea Newman (Los Angeles: Alyson Publications, 2000)
Similar to *Daddy's Roommate*, except it features two lesbian mothers.

How Was I Born?
by Lennart Nilson and Lena Katarina Swanberg
(New York: Dell Publishing, 1996)
This book contains amazing photographs of prenatal development. Gregory loved looking at the pictures of "what I looked like before I was born." It would also be a good book to use with your child if you are pregnant.

See How You Grow: A Lift the Flap Body Book
by Dr. Patricia Pearse and Edwina Riddel
(New York: Barrons' Juveniles, 1991)
This book is for slightly older children (ages five to eight), and includes information about fetal development, infancy, childhood, puberty, and later years. The fold-outs make it fun reading.

Where Do Babies Come From?
by Angela Royston (New York: DK, 2001)
A good introduction to reproduction, with beautiful photography, it would be appropriate for preschool through grade one.

Belly Buttons Are Navels
by Mark Schoen (New York: Prometheus Books, 1990)
An ideal read-aloud book for preschoolers, it introduces the concept that boys and girls are mostly alike, as well as the correct names of genitals.

My Body Is Private
by Linda Walvoord (Morton Grove, IL: Albert Whitman and Co., 1992)
This is the best of the books I've read for preschoolers on sexual abuse prevention. It is gentle, and it models a supportive parent talking about these difficult issues.

Daddy's Roommate
by Michael Willhoite (Los Angeles: Alyson Publications, 1991)
This is a sweet book about two men who love each other. It is appropriate for gay families, but I also shared it with both my children in kindergarten as a way to introduce the topic.

9- TO 12-YEAR-OLDS
There are many good books for girls on puberty; fewer are written for boys. Books are not a substitute for parent education about these issues, but can be a good resource for your child to consult when you've finished sharing.

Girl Stuff: A Survival Guide to Growing Up
by Margaret Blackstone and Elissa Haden Guest
(San Diego: Gulliver Books, 2006)
This book covers a wide range of issues for both the preadolescent and adolescent girl. Liberal parents will like it; conservative parents may find it too wide ranging in its coverage of sexuality topics.

The Period Book
by Karen Gravelle and Debbie Palen
(New York: Walker and Company, 2006)
A funny, light-hearted book with practical information for young girls who are coping with their first menstrual periods.

From Boys to Men: All About Adolescence and You
by Michael Gurian (New York: Price Stern Sloan Publishing, 1999)
A small book on puberty for boys in late elementary school and middle school.

It's Perfectly Normal: Changing Bodies, Sex, and Sexual Health
by Robie Harris (Cambridge, MA: Candlewick Press, 2004)
This is a must-have book for young people beginning puberty. It is illustrated with colorful humorous drawings, and covers everything from bodies, to puberty, to the feelings of adolescence. Good for boys and girls.

It's So Amazing
by Robie Harris (Cambridge, MA: Candlewick Press, 2004)
This book is a little more basic than Harris' *It's Perfectly Normal*, and could be shared with first through fifth graders. Children will appreciate its humor and accessibility, and I appreciated that its illustrations are of diverse people and families.

The Guy's Book: An Owner's Manual
by Mavis Jukes (New York: Crown Publishing Group, 2002)
This is a straightforward book that will take boys from late elementary school through senior year. It is may be too frank for some families.

Period
by JoAnn Loulan and Bonnie Worthen
(Minnetonka, MN: Book Peddlers, 2001)
This "classic" was updated in 2001 with new illustrations. It provides good information about menstruation and pubertal changes. Some parents may find its coverage of some issues too advanced.

Ready, Set, Grow!
by Lynda Madaras (New York: Newmarket Press, 2003)
This is Lynda Madaras' book for younger girls ages eight to eleven, who will appreciate its direct language, simple questions and answers, and cute illustrations.

The "What's Happening to My Body?" Book for Girls
by Lynda Madaras (New York: Newmarket Press, 2007)
This is *the* book that I buy my friends' daughters. It includes chapters on puberty, anatomy, menstruation, sexual intercourse, sexually transmitted diseases, and feelings. A companion workbook, My Body, My Self for Girls, is also available.

The "What's Happening to My Body?" Book for Boys
by Lynda Madaras (New York: Newmarket Press, 2007)
The boys' version. It includes information on puberty, anatomy, body changes, girls, sexually transmitted diseases, and feelings, and a chapter titled, "ejaculation, orgasms, erections, masturbation, and wet dreams." A companion workbook, My Body, My Self for Boys, is also available.

The Care and Keeping of You: The Body Book for Girls
by Valorie Schaefer (Middletown, WI: Pleasant Company, 1998)
A head-to-toe guide on bodies, including basic health and hygiene. It may appeal to conservative parents, as it is one of the few puberty books that do not discuss sex.

What's with My Body? The Girls' Book of Answers to Growing Up, Looking Good, and Feeling Great
by Selene Yeager (New York: Prima Lifestyles, 2002)
Written by one of the health writers at Rodale Press in a question-and-answer format, this book contains advice for preteens and young teen girls on a variety of health issues.

ORGANIZATIONS

The following organizations have resources to help parents talk to their children about sexual issues. You may also wish to contact your local library, pediatrician, health department, Planned Parenthood affiliate, parent-teacher association, church or synagogue, hospital health education department, Girls or Boys Club, Girl or Boy Scouts, school nurse, school guidance counselor, United Way, or YMCA/YWCA for more information.

Advocates for Youth
2000 M Street, NW, Suite 750, Washington, DC 20036
(202) 419-3420; www.advocatesforyouth.org
Advocates for Youth is a national organization concerned with adolescent pregnancy prevention, HIV prevention, and sexual health for teenagers. Call them for their latest publications catalog.

American Academy of Pediatrics
141 Northwest Point Boulevard, Elk Grove Village, IL 60007
(847) 434-4000; www.aap.org
This is the official membership organization for the country's board-certified pediatricians. Their Web site contains breaking news related to pediatrics. They have pamphlets for parents and an excellent guide to school health.

American Library Association
50 East Huron Street, Chicago, IL 60611
(800) 545-2433; www.ala.org
The American Library Association offers excellent reading lists for children and adults, including one on sexuality, and is also a resource for fighting censorship of materials.

American Medical Association
515 N. State Street, Chicago, IL 60610
(800) 621-8335; www.ama-assn.org
This is the country's leading medical organization. Their division of adolescent health provides excellent materials for parents of adolescents. Their Web site is easily searchable for medical information.

American Psychological Association
750 First Street, NE, Washington, DC 20002
(800) 374-2721; www.apa.org
Contact the American Psychological Association for referrals to trained, certified psychologists.

Center for Disease Control and Prevention—National Prevention Information Network
1600 Clifton Road, Atlanta, GA 30333
(404) 639-3311; www.cdc.gov

This organization is run by the federal government and includes extensive information about HIV prevention. It is also a good place to call if you want more information about a breaking news story on AIDS.

Center for Media Literacy
23852 Pacific Coast Highway, #472, Malibu, CA 90265
(310) 456-1225; www.medialit.org
This organization has excellent materials on sharing television with your children.

Child Welfare League of America
2345 Crystal Drive, Suite 250, Arlington, VA 22202
(703) 412-2400; www.cwla.org
The Child Welfare League of America has excellent information on foster care, teenage pregnancy, and adoption.

ETR Associates
4 Carbonero Way, Scotts Valley, CA 95066
(831) 438-4060; www.etr.org
Write or call for their latest catalog of publications. ETR also publishes "When Sex Is the Subject," a good guide for parents and teachers.

Girls, Incorporated
120 Wall Street, 3rd Floor, New York, NY 10005-3902
(800) 374-4475; www.girlsinc.org
Girls Incorporated has clubs around the country for girls from elementary school through adolescence. I especially recommend their mother/daughter programs on puberty preparation.

Mothers' Voices
150 West Flagler Street, Suite 1820, Miami, FL 33130
(305) 347-5467; www.mothersvoices.org
This organization promotes mothers' involvement in the HIV/AIDS epidemic. They have publications for parents and resource information on their Web site.

National Council for Adoption
225 North Washington Street, Alexandria, VA 222314
(703) 299-6633; www.adoptioncouncil.org
This organization provides resources for adoptive parents.

The next two organizations can help you if you suspect or are dealing with child sexual abuse.

National Center on Child Abuse and Neglect
P.O. Box 1182, Washington, DC 20013-1182
(202) 205-8586; www.casanet.org/library/abuse/nrccan.htm

National Resource Center on Child Sex Abuse
Central Bank Building, Huntsville, AL 35801
(800) KIDS-006

National Campaign to Prevent Teen Pregnancy
1776 Massachusetts Avenue, NW, Suite 200, Washington, DC 20036
(202) 478-8500; www.teenpregnancy.org
This organization is dedicated to reducing teenage pregnancies in the United States. They have resources on how parents can become involved.

National Eating Disorders Association
603 Stewart St., Suite 803, Seattle, WA 98101
(206) 382-3587; www.nationaleatingdisorders.org
This organization can help you if you suspect your son or daughter is developing an eating disorder.

National Education Association—Health Information Network
1201 16th Street, NW, Suite 216, Washington, DC 20036-3290
(202) 833-4000; www.nea.org/hin
The NEA is one of the two major teacher unions in the United States. Their Health Information Network has materials on HIV prevention, school health education, and sexuality education.

National Information Center for Children and Youth with Disabilities
P.O. Box 1492, Washington, DC 20013-1492
(800) 695-0285; www.nichcy.org
A good source of material for families that have children with developmental disabilities. They have resources on teaching children with disabilities about sexuality.

National Maternal and Child Health Clearinghouse
Health Resources and Services Administration
Parklawn Building, 5600 Fishers Lane, Rockville, MD 20857
(888) 275-4772; www.ask.hrsa.gov
This clearinghouse, funded by the federal government, has good materials on pregnancy, child health, and small children.

National Organization on Adolescent Pregnancy, Parenting, and Prevention
509 2nd Street NE, Suite 350, Washington, DC 20037
(202) 547-8814; www.healthyteennetwork.org
This organization provides resources to educators and providers on adolescent pregnancy, parenting, and prevention. Its newsletter may be helpful to parents with older children.

Parents, Families, and Friends of Lesbians and Gays
1726 M Street, NW, Suite 400, Washington, DC 20036
(202) 467-8180; www.pflag.org
This is a support group for parents and families of children who are gay and lesbian. Although most of their materials are for parents of teenage and adult children, some of their materials address gender issues in earlier childhood. The pamphlet "Our Trans Children" may be helpful to parents wondering about gender orientation.

Planned Parenthood Federation of America
434 West 33rd Street, New York, NY 10001
(212) 245-1845; www.plannedparenthood.org
PPFA is the largest network of providers of family planning services. They publish excellent materials for parents, and their affiliates provide local education programs for parents in communities.

Prevent Child Abuse America
500 N. Michigan Avenue, Suite 200, Chicago, IL 60611
(312) 663-3520; www.preventchildabuse.org

Rape, Abuse, and Incest National Network
2000 L Street NW, Suite 400S, Washington, DC 20036
(800) 656-HOPE; www.rainn.org
This organization can help you deal with cases of rape, sexual abuse, and incest.

Religious Institute on Sexual Morality, Justice, and Healing
21 Charles Street, Suite 140, Westport, CT 06880
(203) 222-0055; www.religiousinstitute.org
This is the organization that I cofounded in 2001. It includes resources for faith communities on sexuality, including lists of sexuality education curricula developed to be used in religious institutions.

Search Institute
The Banks Building, 615 First Avenue NE, Suite 125, Minneapolis, MN 55413
(612) 376-8955; www.search-institute.org

The Search Institute promotes healthy children, youth, and communities. The section on "Information for Families" includes evidence-based reports on what parents need to succeed.

SIECUS
90 John Street, Suite 704, New York, NY 10038
(212) 819-9770; www.siecus.org
SIECUS is the organization that I used to head. It has extensive booklets, bibliographies, and teacher materials, and a Web site for parents.

Society for Adolescent Medicine
1916 NW Copper Oaks Circle, Blue Springs, MO 64015
(816) 224-8010; www.adolescenthealth.org
Call them for a list of doctors specializing in adolescence in your area.

Teach-A-Body Dolls
7 Dons Drive, Mission, TX 78572
(956) 581-9959; email: TABDOLL@aol.com

HOTLINES
These 800 numbers are staffed by trained counselors and educators who can help answer your questions and make referrals to local care providers.

American Dietetic Association Nutrition Hotline . . . . . . (800) 366-1655

National STD Hotline . . . . (800) 227-8922

Domestic Violence Hotline . . . . (800) 799-7233

National AIDS Hotline . . . . English (800) 342-AIDS

. . . . Spanish (800) 344-7432

. . . . TTY (800) 243-7889

National Eating Disorders Screening Program . . . . (800) 969-6642

National Child Abuse Hotline . . . . (800) 4A-CHILD

National Gay and Lesbian Hotline . . . . (888) 843-4564

Planned Parenthood Federation of America . . . . (800) 230-7526

References

Alan Guttmacher Institute. *Sex and America's Teenagers*. New York: The Alan Guttmacher Institute, 1994.

American Academy of Pediatrics. Circumcision Policy Statement (RE9850), *Pediatrics*, No. 103:3, (1999): 686–693.

American Academy of Pediatrics. "New AAP Circumcision Policy Press," March 1, 1999.

"Assisted Reproductive Technology Surveillance 2000." http://www.cdc.gov/reproductivehealth/ART00

Beal, Carole R. *Boys and Girls: The Development of Gender Roles*. New York: McGraw Hill, 1994.

Bernstein, Anne C. *Flight of the Stork*. Indiana: Perspectives Press, 1994.

Bryant, Anne. "Hostile Hallways: The AAUW Survey on Sexual Harassment in American Schools," *Journal of School Health* 63, No. 8 (October 1993): 355–57.

Bullough, Vern L., and Bonnie Bullough. *Human Sexuality: An Encyclopedia*. New York: Garland Press, 1994.

Bureau of Labor Statistics. *Labor Force Statistics from the Current Population Survey*, 1997 (from Web site http://cpsinfo@bls.gov).

Carnegie Corporation of New York. *Great Transitions*. New York: Carnegie Corporation of New York, 1996.

Centers for Disease Control. *Hepatitis B Vaccine Pamphlet*. Atlanta: Centers for Disease Control and Prevention, 1996.

Centers for Disease Control and Prevention. "Youth Risk Behavior Survey, 2001." http://www.cdc.gov

Clapp, Steve, et. al. *Faith Matters: Teenagers, Religion and Sexuality*. IN: Lifequest, 2003.

Coles, Robert, and Geoffrey Stokes. *Sex and the American Teenager*. New York: Harper and Row Publishers, 1985.

The Commonwealth Fund Survey of the Health of Adolescent Girls. New York: The Commonwealth Fund, November 1997.

Dickey, Nancy W. "To Circumcise or Not to Circumcise," 2002, http://www.medem.com.

Early Childhood Sexuality Education Task Force. *Right From the Start: Guidelines For Sexuality Issues, Birth to Five Years.* New York: SIECUS, 1995.

Elias, Marilyn. "When 'Friends' Talk, Teens Listen," *USA Today,* November 3, 2003.

Erikson, E. H. *Childhood and Society.* New York: Norton, 1950.

——. *Identity: Youth and Crisis.* New York: Norton, 1968.

Finkelhor, D. *A Sourcebook on Child Sexual Abuse.* California: Sage Publications, 1986.

Friedrich, William N., Jennifer Fisher, Daniel Broughton, Margaret Houston, and Constance R. Shafran. "Normative Sexual Behavior in Children: A Contemporary Sample," *Pediatrics* Electronic Page 101, No. 4 (April 1998): e9.

General Social Survey. Cumulative File 1972–1994. Chicago: National Opinion Research Center, 1998.

GetNetWise. "Tools for Families." http://kids.getnetwise.org/tools.

Girls Inc. *Re-Casting TV: Girls' Views.* New York: Girls Inc., 1995.

Gordon, Sol. *You.* New York: Quadrangle/The New York Times Book Co., Inc., 1975.

Haffner, Debra, ed. *Facing Facts: Sexual Health for America's Adolescents.* New York: SIECUS, 1995.

Herman-Giddens, Marcia. "Secondary Sexual Characteristics and Menses in Young Girls Seen in Office Practice: A Study from the Pediatric Research in Office Settings Network," *Pediatrics* 99, No. 4 (1997): 505–12.

Kaiser Family Foundation. "See No Evil: How Internet Filters Affect the Search for Online Health Information." Executive Summary, December 2002.

Kelly, Maureen. "Ten Tips for Raising Kids with a Healthy Body Image." http://www.advocatesforyouth.org/parents/experts/kelly.htm.

Komar, Miriam. *Communicating with the Adopted Child.* New York: Walker Publishing Co., 1991.

Lapinski, Emaline. "To the editor," *New York Times,* October 26, 2003.

Leitenberg, Harold, Evan Greenwald, and Matthew Tarran. "The Relation Between Sexual Activity Among Children During Preadolescence and/or Early Adolescence and Sexual Behavior and Sexual Adjustment in Young Adulthood," *Archives of Sexual Behavior* 18, No. 4 (1989): 299–313.

Levine, M. P., L. Smolak, H. Hayden. "The Relation of Sociocultural Factors to Eating Attitudes and Behaviors Among Middle School Girls," *Journal of Early Adolescence 4,* No. 4: 471–90.

Louis Harris and Associates, Inc. *In Their Own Words: Adolescent Girls Discuss Health and Health Care Issues.* New York: The Commonwealth Fund, March 1997.

Lung, C. T., and D. Daro. *Current Trends in Child Abuse Reporting and Fatalities: The Results of the 1995 Annual Fifty State Survey.* Chicago: National Committee for the Prevention of Child Abuse, 1996.

Magin, L. J. *Child Safety in the Information Society.* Arlington: National Center for Missing and Exploited Children, 1994.

Maloney, M. J. "Dieting Behaviors and Eating Attitudes in Children," Pediatrics 84 (1989): 482-87.

Masters, William H., Virginia E. Johnson, and Robert C. Kolodny. *Human Sexuality.* New York: HarperCollins Publishers, Inc., 1992.

Media and Values Magazine, No. 46 (Spring 1989).

Miller, B. C. *Families Matter: A Research Synthesis of Family Influences on Adolescent Pregnancy.* Washington, DC: National Campaign to Prevent Teen Pregnancy, 1998.

Miner, Barbara. "Internet Filtering: Beware the Cybercensors," Rethinking Schools (Summer 1998).

Money, John. *Principles of Developmental Sexology.* New York: The Continuum Publishing Group, 1997.

National Campaign to Prevent Teen Pregnancy. "14 and Younger: The Sexual Behavior of Young Adolescents." http://www.teenpregnancy.org

National Guidelines Task Force. *Guidelines for Comprehensive Sexuality Education, Kindergarten–Twelfth Grade,* 2nd ed. New York: SIECUS, 1996.

Neinstein, Lawrence S. *Adolescent Health Care: A Practical Guide.* Baltimore: Urban and Schwarzenberg, 1984.

NetSmartz Parents. "Online Risks." http://www.netsmartz.org/parents/home/rskinfo.html.

Papilia, Diane E., and Sally Wendkos Olds. *World: Infancy through Adolescence.* New York: McGraw Hill, 1993.

Remafedi, G., et al. "Demography of Sexual Orientation in Adolescents," *Pediatrics* 89, No. 4 (1992): 714–21.

Resnick, M. D. "Protecting Adolescents from Harm," JAMA 278, No. 10 (September 10, 1997): 823–32.

Rideout, Victoria, and Tina Hoff. *Sex, Kids, and the Family Hour: A Three-Party Study of Sexual Context on Television.* California: Children Now and the Henry J. Kaiser Family Foundation, 1996.

Schaeffer, Judith, and Christina Linstrom. *How to Raise an Adopted Child.* New York: Crown Publishers, 1989.

Smolak, L., and M. Levine. "Ten Things Parents Can Do to Help Prevent Eating Disorders in Your Children," *National Eating Disorders Organization Newsletter* XVII, Issue 3 (July–September 1994).

Stansell-Gamm, Martha. "There's One More Talk You Need to Have," *Newsweek*, September 15, 2003.

Tartaglione, Ralph. "Kids Online: What Parents Can Do to Protect Their Children from Cyberspace," *SIECUS Report* 25, No. 1.

Task Force on Circumcision. "Report of the Task Force:" *Pediatrics* 84, No. 2 (August 1989): 388–91.

U.S. Census Bureau, "Profile of Selected Economic Characteristics." http://factfinder.census.gov.

U.S. Department of Health and Human Services. "More Americans of All Ages Are Overweight," press release, March 6, 1997.

Walker, Richard. *The Family Guide to Sex and Relationships.* New York: Simon and Schuster, 1996.

Wallerstein, Judith S. "Children of Divorce: Report of a Ten-Year Follow-Up on Early Latency–Age Children," *American Journal of Orthopsychiatry* 57, No. 2: 199–211.

Wilson, Pamela. *When Sex Is the Subject: Attitudes and Answers for Young Children.* California: Network Publications, 1991.

Index